A LITTLE THEORY
OF ALMOST
ABSOLUTELY EVERYTHING
CONSCIOUSNESS AND THE COSMOS

Crystal Love

Published in the UK 2023 by Spiral Press (London)

Copyright © Crystal Love 2023

Crystal Love has asserted her right under the Copyright, Designs and Patents Act, 1988, to be identified as the author of this work.

All rights reserved. No part of this book may be reproduced, stored in a retrieved system or transmitted, in any form or by any means, electronic, mechanical, scanning, photocopying, recording or otherwise, without the prior permission of the author and publisher.

Paperback ISBN: 978-1-7394324-0-9
eBook ISBN: 978-1-7394324-1-6

Cover design and typeset by SpiffingCovers.com

'Mind is the forerunner of all things
Mind is foremost
Of mind, are all things made'

The Dhammapada. Buddhist text. Verse I. 3rd Century BC

Contents

AUTHOR'S NOTE: 7
INTRODUCTION: 9

THE UNIVERSAL MIND. 17
    The Energy God. 17
    Before Creation. 28
    In the Beginning. 33
    The Big Bang. 38
    Astrology. 40

THE HUMAN MIND. 61
    The Invisible Universe. 61
    The Aura - Soul Within 69
    Auric Healing. 72
    Conscious and Unconscious. 78
    Left and Right Brain 80

MIND OVER MATTER. 85
    Hypnosis. 85
    Altered States. 90
    Sound, Music and Sacred Dance. 104
    Meditation, Trance and
    Transcendental Experience. 109
    Miracles and Magic. 113
    The Power of Positive Thinking. 116

FUTURE MIND. 122
    Space-Time and the Sixth Sense. 122
    Telepathy & ESP 129
    Prediction and Prophecy. 137
    Dreams and Premonitions. 141

EVERYTHING IS MIND 147
- The Out-Of-Body-Experience 147
- The Near Death Experience. (NDE) 156
- Death and Resurrection 165
- Visions of Paradise. 168
- Ghosts, Mediums and Spirit Guides 172
- Mediums 183
- Spirit Guides 187

IMMORTAL MIND 193
- The Law of Cause and Effect 193
- The Law of Karma 194
- The Theory of Reincarnation 206
- Past Life Regression 215
- Return to the Source. 226

BIBLIOGRAPHY. 229
INDEX 233
ABOUT THE AUTHOR. 247

AUTHOR'S NOTE:

Science tells us, quite unequivocally, that everything in the Universe, including you and me, is made from pure Energy. Furthermore, this Energy is infinite, eternal and can never be destroyed. It would stand to reason then, that there is something within ourselves that is also infinite, eternal and indestructible. Is this what the ancients referred to as 'the soul', and does the existence of this soul explain so-called supernatural or mystical experiences?

Whilst science concentrates on trying to find a 'Theory of Everything' and attempts to understand the Cosmos by measuring physical matter - be it enormous Black Holes, or minute subatomic particles - consciousness, the very thing that makes either possible, is dismissed as being unworthy of serious investigation.

However, given the growing reports of experiences of 'non-ordinary reality' that defy the limited concept of consciousness as currently held by scientists, medics and psychologists alike, and that appear to defy the very laws of time and space as we understand them, I believe the subject calls for further objective investigation.

Using hitherto ignored or unpublicised scientific data, theory and evidence, I hope to be able to prove to you in the pages of this book that the existence of these experiences is scientifically feasible. These experiences, the validity of which are often questioned and rejected by both formal religious doctrine and scientific understanding, have nonetheless become nothing short of a pandemic social phenomena within the last few decades of our collective history.

Furthermore, these experiences are being reported trans-globally at an ever-increasing rate. They also appear to exist entirely independently of any one particular religion, culture or belief system, and regardless of race, creed, colour, age or sex. These experiences challenge the very nature of traditional Western materialistic science and religious dogma.

Ultimately, what I seek in this work are universal truths that are free of political, religious or scientific dogma, and which speak a language of their own. Based on extensive research and inspired by my own experiences which began as a child and which continue to this day, I later became increasingly determined to find a logical framework of reference for these experiences - for which neither my parents, school, society, scientific or religious teaching were able to prepare me.

Since no one Western religion, philosophy or scientific doctrine seemed to encompass or refer to these experiences in any meaningful way, and indeed, most were openly hostile on such matters, I studiously began my own in depth research for the evidence that my experiences required of me.

These experience, having no root to speak of in my own cultural frame of reference at the time, begged me to find a thread that would lead me to the 'Holy Grail' a 'Theory of Everything' that would unite both religion and science in a harmonious cohesive oneness, whilst at the same time incorporating the reality of the individual mystical experience in terms acceptable to both.

This book is an attempt to reveal those findings.

INTRODUCTION:

Even though humankind has begun to explore the farthest reaches of outer space and developed the material technology to land on the Moon relatively recently in our shared history, inner space, the human mind, our incredible consciousness, remains relatively uncharted, and its many great mysteries remain unsolved.

And yet the remarkable personal testimonies and experiences of literally millions of people worldwide are demanding a change in our modern western belief system. The increasing reports of personal experiences of a mystical, supernatural, transcendental or spiritual nature, such as Telepathy and ESP, the Out-of-Body, Near-Death-Experience, and Past Life Memory, have become all too numerous. Far too numerous to dismiss or ignore. As the evidence for these phenomena grows, many of us are asking the inevitable questions about ourselves, and our place in the universe.

Our culture, our lives, and our individual thinking have always been influenced and affected by the many visible and invisible forces that operate within society as a whole at any one time. Moreover, our understanding of the nature and meaning of the Universe and the laws that operate within it are defined, to a large extent, by these social genres and the norm can be defined, and redefined, by society as a whole at any given point in its history.

Our current modern Western belief system has understandably been influenced by the religion, politics and science of our time, and the collective Western concepts of the universe, of God, soul, judgement, afterlife, of heaven and hell, have all been influenced to a large extent by Judeo-Christian concepts.

Today, however, after 2,000 years of Christianity, many

Churches are empty of their congregations and large numbers have now fallen into disuse or disrepair. The Church of England is also simultaneously beset with internal sex scandals, and maintained its selective, stubborn and sexist refusal to ordain women priests until 1992, when the General Synod voted in their favour by a margin of just two votes.

The Church of England, now worth approximately £3 billion in land, property, and stocks & shares (which, until recently included shares in the armaments industry) has dominated our religious views for nearly five hundred years since Henry VIII divorced himself and England from Rome to marry his mistress, Anne Boleyn.

In the Vatican, the Pope too, has his own cross to bear. In the first of two controversial books, 'Hitler's Pope', author John Cornwell (Senior Research Fellow at Jesus College, Cambridge), documents historical evidence of the Nazi sympathies of wartime Pope Pius XII - Eugenio Pacelli. According to the author, Pius XII was an anti-Semite who coerced on the side of Fascism, entering into pre-war concordats with Hitler, which later ensured financial advantages for the Vatican in return for non-intervention in the horrors of the holocaust. These concordats later helped usher the Fuhrer to power and minimise Catholic opposition to the Final Solution. (It should be remembered however, that although Pius XII did little or nothing to help save the Jews, many ordinary Catholic priests and nuns risked their own lives to help save the gravely endangered Jewish population of Europe.)

As Pontiff, surely Pius XII must have been familiar with Biblical literature? On that basis, it is hard to believe that he could have missed the passage in the New Testament in which Christ (himself born a Jew) commands us to 'Love thy neighbour as thyself.' (Mathew: (5:43)). It is even harder to believe that a man who was so obviously flawed as a Christian managed to become Pontiff, head of the Roman Catholic Church, and Christ's alleged 'representative' on Earth.

Neither is Pius XII's tolerance of fascism and ethnic cleansing unique in Papal history. From Pope Urban II - who preached the first crusade against Islam in 1095 - to the latter day Popes who have openly supported such fascist dictators as Franco and Mussolini - successive Pontiffs have demonstrated an absolute spiritual disregard for essential humanity and Christian teaching.

The Vatican, already seriously disturbed and discredited by John Cornwell's accusations, comes under further attack in the second book. In 'Gone with the Wind in the Vatican', the writers - a pseudonymous group who call themselves 'I Millenari' - level accusations of corruption, Freemasonry, sexual scandal and devil worship within the Papal Court, The Curia. Monsignor Marinelli, who worked in the Curia for 35 years, and whose name is an anagram of I Millenari, has now admitted taking part.

Despite Christ's own views on material wealth, best expressed in the following Biblical quotation: 'It is easier for a camel to pass through the eye of a needle, than for a rich man to enter the gates of Heaven' (Mathew 19.24), the Vatican remains one of the wealthiest religious institutions in the modern world.

In 1996, with scandals of its own to contend with, the Church of England issued a report 'In Search of Faith' in which it stated that although only less than 15% of the population of the UK still go to Church, over 75% nonetheless still believed in God, and even more remarkably, in Reincarnation!

The Church - which might do well to contemplate that other well-known Biblical maxim 'Caste out the mote in your own eye, before you caste out the mote in your brother's eye.' (Mathew 7:3) - is now blaming astrology, alternative, new-age, neo-pagan and indigenous mystical folk religions for the decline of the Church itself and society as a whole! It warns us against this return to 'darkness and superstition' and the dangers of an eclectic 'pick and mix' philosophy.

However, before the Romans came to Britain in the

fourth century A.D., bringing formal Christian doctrine with them, Britain was a Pagan country that was rich in mystical nature religions and traditions. Stonehenge, a complex astrological observatory, solar temple and place of religious worship, was built thousands of years before the Romans converted the early Britons to Christianity. This they did with remarkable ease, by superimposing Christian churches on pagan holy sites, and incorporating Christian festivals into pagan ones. Easter and Christmas, for instance, were arranged to coincide with the Pagan Holy Days - the Spring Equinox and Winter Solstice respectively.

Although the male dominated Church and the new religion eventually suppressed the old mystical nature religions with fierce determination, culminating in the burning of so called 'heretics' and 'witches' in the Middle Ages, many old pagan customs, like Halloween survive to this day.

We are now witnessing the decline of the Church and the rise of 'New Age' religions that readily incorporate the magical and mystical. The evolution of a 'New Age' religious culture furthermore, appears to fully embrace the concept of personal empowerment through the mystical and transcendental experience. In this sense, new age philosophies, beliefs and customs appear to resemble early mystical nature and pagan religious traditions in many respects.

What is mysticism? The word 'mystic' originally comes from the Greek word for 'mystery' - meaning 'the unexplainable' - whose own root word means quite simply, 'with closed eyes. According to mystic philosophy, one cannot gain a full understanding of God, the nature of the Universe, or indeed the 'self' by external or physical means alone. By 'closing the eyes' and shutting out the external physical world of the five senses through introspection, meditation or prayer, a divine union with God, vision or revelation may be experienced by the individual.

Meditation, introspection and prayer are common to all religions, and yet mysticism is not a religion because it has no hierarchy, no rules, no regulations, and no sacred texts. In mysticism the central character is yourself and you are, in your highest sense, both the worshipped and the worshipper, both temple and congregation. The mystical journey is, by its very nature, a journey into the self, and a journey ultimately made alone.

However, the mystical and supernatural may be experienced spontaneously by those with no religious affiliations at all atheists and agnostics alike. Apparently regardless of all belief and outward form, the mystical experience appears to exist and function as an internal vision or revelation within the mind or the 'soul' of the self, and not as an externalised sensual experience.

The mystic tradition itself pre-dates organised religion and is ultimately the very foundation of all religions, whether they are pagan, primal or otherwise and mystical revelation can be clearly seen to be at the heart of them all. This includes the Eastern philosophies of Hinduism and Buddhism as well as the three Western monotheistic religions, Judaism, Christianity and Islam, the latter of which were founded on the mystical revelations of their own respective teachers or prophets Buddha, Moses, Jesus Christ and Mohammed.

Although the three great monotheistic religions which originated in the Middle East - Judaism, Christianity and Islam - are based on the individual vision, revelation, and mystical experience of these divine messengers or prophets, it has been the case throughout recorded history that some of these self same religions have also been largely responsible for the repression of individual empowerment through the internalised individual experience of personal revelation.

However innocent and natural this primal human desire for higher spiritual states of consciousness may be, and it is obviously a collective social phenomena from the beginning, the art, knowledge and practice of spiritual mysticism has found no safe haven in our relatively modern male-dominated monotheistic religions.

In the words of Carl Jung however: 'So long as religion is only faith and outward form, and the religious function is not experienced in our own souls, nothing of any importance has happened. It has yet to be understood that the 'mysterium magnum' is not only an actuality, but is first and foremost rooted in the human psyche. The man who does not know this from his own experience may be a most learned theologian, but he has no idea of religion and still less of education.'

The philosophy of mysticism and individual revelation however is a unifying philosophy - as it recognizes the right of every individual to this state of harmony, bliss and union with the divine, which can only be experienced by the individual as an internalised phenomenon.

As well as suffering dwindling congregations, both the Protestant and Catholic Churches are now in further turmoil over their own accepted beliefs and philosophies. These were established at the Council of Nicea in Constantinople in 325 A.D., when the council members were called to 'define' the 'real' meaning of the scriptures. However, many of the original scriptures were strangely omitted from the New Testament at this time, including the Gospel of Enoch, which remain hidden in the vaults of the Vatican to this day.

Furthermore, in August 1999 - days after a leading Jesuit magazine redefined concepts of Hell - the Pope declared that Heaven was not 'a place above the clouds where angels play harps', but a state of being after death. 'The Heaven in which we will find ourselves is neither an abstraction, nor a physical place amongst the clouds', the Pontiff told pilgrims in St. Peter's Square.

Just as 'hell' was separation from God, a state of being in which those who had consistently rejected God were condemned to banishment from the heavenly presence, so Paradise was 'close communion and full intimacy with God', God furthermore, was not 'an old man with a white beard' - but rather a supreme being with both masculine and feminine aspects.

Now, however, after nearly 2,000 years of threatening it's congregations with 'hellfire and brimstone', and threatening any freethinking individual with the charge of 'heresy' if they should dare challenge the doctrine of the Church, the Church itself is beginning to adopt the self same views and philosophies which have been at the very heart of nature, pagan, primal and Eastern religions all along.

In the meantime, the Church of England is arguing amongst itself as to the real nature of the Resurrection, and in the Archbishop of Canterbury's Millennium message, Dr George Carey says we 'cannot know' Jesus was resurrected from the dead.' It goes against all human experience and our first instinct is incredulity'.

The Virgin birth too, is now coming under sceptical scrutiny and whilst the Church concentrates on re-defining its own beliefs, science has taken an altogether different viewpoint in its own short history, culminating in the 'God is dead' philosophy of scientific materialism so popularised during the sixties.

Today, however, we are witnessing a spiritual revival and revolution which is happening at the grassroots level of society as a whole, and which appears to be a culture in its own right, regardless of the views of either scientific or traditional religious thought and dogma. Astrology, healing, telepathy and ESP., the out of body & near death experience, as well as reincarnation and past life regression, have become bridges to a new religion.

Are we witnessing the birth of a new mystical philosophy evolved by a generation that has perhaps grown tired of listening to a Church that neither understands, or practises what it preaches, and a science which has denied the sacred mystery and magic of the world?

Disillusioned and dissatisfied with both science and formal religion, people around the world are now looking to define their beliefs, spiritual needs and experiences in some other way. Ironically however, it is modern science with its latest discoveries and theories in Superstring and Quantum Entanglement that is now - albeit inadvertently - offering verifiable scientific evidence and information on the nature of the Universe and the validity of the mystical and supernatural that I believe, will soon come to confirm the unity of all things and the amazing untapped power and potential of our incredible human consciousness.

Chapter One.
THE UNIVERSAL MIND.

The Energy God.

'The Primal Mind, which is life and light, being bisexual, gave birth to the Mind of the Cosmos. The Primal Mind is ever unmoving, eternal and changeless and contains within it this Cosmic Mind which is imperceptible to the senses. The cosmos, which senses perceive, is a copy and image of this eternal Cosmic Mind, like a reflection in a mirror'.
                    The Hermetica. Egyptian Text. 3,000 BC.

Since the dawn of time, humanity has wondered about the meaning of life and pondered the nature of the universe. To this day we are still wondering and debating. The universe is nobody's fool however, and does not give up her secrets easily. Alone in the vastness of this magnificent star-filled cosmos, resplendent on a uniquely lush blue orb, we are all spiralling helplessly, relentlessly and irrevocably through space and time.

Where are we going? Why are we here? Is there a meaning and purpose to life? Where did the universe come from? Is there a God? What happens to us when we die?

Throughout the ages, many diverse religions, philosophies, cosmological and scientific theories have come into being in an attempt to answer these questions. Diverse as these multiple theories may be however, and there are as many as there are people in the world, they can be roughly divided into two opposing camps.

The first camp - that of religion, spirituality and mysticism - has collectively, throughout history, contained three constantly recurring themes:

1) That the universe is a deliberate manifestation of a divine mind or intelligence - an act of God in other words.

2) That all creatures have an eternal soul which survives physical death and which will be called to account for its deeds by God on 'Judgment Day'.

3) That some people are naturally gifted with the ability to intercede with, or contact these higher spiritual realms or dimensions - usually through mystical vision or revelation. This would include mystics, holy men, shaman and prophets.

This first camp also believes in the possibilities of higher realities, or dimensions, and an ultimate state of spiritual bliss, heaven or nirvana. They also collectively recognize and have a reverence for the 'Angelic' or spiritual planes of existence. The Angel Gabriel for instance, is quintessential in the Old and New Testaments, as well as the Quran. Sometimes this camp also incorporates a belief in reincarnation, depending entirely on which spiritual system the individual camp member follows.

The second camp of course, is that of the scientific materialists and atheists who also have three main, but contradictory beliefs:

1) There is no God. The universe and life on Earth, with all its intricate and detailed diversity, 'just happened' - atoms accidentally bumped into each other and hey presto! (According to one theorist, that's about as likely as a jumbo jet in bits on the tarmac being assembled by a strong wind!)

2) We have no soul - we are just physical forms. According to this camp, you live, you die, and that's it. There is nothing more. There is no God, no heaven, no reincarnation, no life after death. No nothing!

3) The 'supernatural' is unscientific and serves no useful purpose in a practical world. According to this camp, the other camp is basically 'out to lunch with the fairies', and their 'visions' and 'revelations' are nothing more than delusion, hallucination or wishful thinking. This camp does not believe in anything unless it is physically scientifically

verifiable or at the very least, has some logical or scientific explanation.

Of course, common sense tells us that both contradictory philosophies cannot be right - there either is, or is not, a supreme creative intelligence in the universe - with all its implications. One camp, therefore, has a logic that is fatally flawed.

Now, and this is the eternal billion dollar question - which one of these two camps is right? Before we can embark on any meaningful debate on the subject, however, we should firstly consider and contemplate that ancient maxim 'As above, so below', for should we be in any doubt as to our relationship with the universe from the start, we can be reminded quite simply, that the laws which apply in the universe apply to both great and small alike.

The macrocosm and the microcosm, the universe and the atom, are identical in principle (both essentially consisting of orbiting bodies around a central sphere) and since we live in one and are composed entirely of the other, there is no reason to suppose that we are somehow separate from the same principles that apply at the very heart of both. We are all, whether great or small, bound together in the same web of creation and like it or not, we obey the same laws, follow the same universal principles and are all inexorably and eternally linked.

Furthermore, we cannot separate any one part of the universe and exclude it from the universe itself any more than you could take any one cell from your body and deny it equal status and participation in your own being. Whether that cell is far away on the end of your big toe, or right next to your line of vision on your nose, it is still part of a comprehensive whole that makes up your totality. It is the same with the body of the universe, and despite its enormity, every part of it is part and parcel of its entirety. And that includes you!

If somewhere in one's own mind the concept of

separateness should try to take hold - that is, to believe that any one part of creation is not intrinsically bound and connected to any other part of creation and that somehow, the laws do not apply to 'this' or 'that' or 'him' or 'her' - be rid of it now - because all of creation is inexorably linked, bound together in a wholeness or 'Oneness'.

We cannot, for instance, separate out the Sun from the Moon and suggest one or the other is not of the same world or subject to the same laws, or even immune from the effects of each upon the other. At the same time, neither are you separate from, greater or lesser than, any other part of creation. You are part and parcel of it by its own, and your very own nature, and from that there is no escape.

However, that is not to say that we do not or may not experience a sense of separateness from time to time. Indeed, this may be at the heart of many a human malady. Our sense of separateness however, is the 'Maya', the illusion of our existence as separate from the whole. Like the blind men who tried to describe an elephant by holding its ear, tail or leg, we may never know anything about the universe until we try to see it in its totality.

So - back to our billion-dollar question - which camp is right? To try and find the answer to that question in any meaningful way, the logic of one must obviously stand the test of scrutiny, and the theories of the other must be firmly, but politely eliminated.

Before we can investigate the many complex and varied implications of the first option however, (the pro-God camp) let us firstly see if the second one (the no-God camp) fits into the world model and follows the same laws and principles as the rest of the Universe. For, as we have already noted, the laws that operate in the Universe apply without exception, and unless the 'accidental universe' meets this criterion, it will automatically and immediately disqualify itself as a reasonable and logical hypothesis.

Now, the universe works on certain mathematical

principles and laws that are part of the very fabric and structure of creation, and one of these is Newton's Third Law of Motion, otherwise known as the Law of Cause and Effect. The Third Law of Motion simply states that: 'every action has an equal and opposite reaction' and simply put, if there is an effect, then there is, without doubt, a proceeding cause.

Following the logic of the Third Law of Motion then, we can assume that if the Universe was created at the moment of the Big Bang, there must have been a causative factor that preceded it. The Universe itself could simply not have come into being without a causative factor, and to suggest that it could do so would simply be contrary to the very laws and nature of its own being.

According to the principles of the Law of Cause and Effect, we would have to conclude that the Universe would have to have had a causative factor, that is, something that both proceeded and caused the Big Bang to happen in the first place. This alone is sufficient reason to doubt the concept of a Universe with no proceeding cause, and if this Law did not apply throughout the universe, I can only see that the universe itself would be without motion, cause or effect, and that matter - if it existed at all - would simply stagnate.

Everything in the known universe is subject to these same universal laws and the very same laws preclude the idea of a universe that had no apparent cause to precede the effects of its own visible manifestation. There must have been, by the very nature of the universe and all the laws that govern it, a proceeding cause. There can be no other explanation. The Universe, operating and abiding by its own laws, and by its very nature must have had a causative factor. This causative factor we collectively refer to in spiritual terms as 'God'.

Now, having failed the first test, that of conforming absolutely to universal laws, the logic of camp number two, the 'no-God' (no causative-factor) camp, has been quickly and quietly eliminated. We must now scrutinise the logic of camp number one, the pro-causative factor (God) camp, and see if their theories about God and the universe stand up to scrutiny!

However, two things should be clarified before we can even attempt to gain some understanding of who or what this 'God', this 'causative factor' could be. If we assume that this God is the sole causative creative factor that gave rise to the materialised universe, we can immediately and irrevocably discount the theory that this God could in any way be human, or possessed of human form. Firstly, because the universe itself is some 14 billion years old, and homo-sapiens were not even present at the time, having only appeared on Earth relatively recently, and secondly, because no one human being has the power to individually create a whole universe.

The popularised Western concept of a milky white God with a cute fluffy beard floating somewhere around Heaven therefore, is a rather absurd concept which no doubt arose from the misunderstanding and misinterpretation of the Biblical statement which appears in Genesis, the first book of the Bible and which states that 'God made man in his own image'. (Genesis 1:26)

As the creator, or creative force of the entire universe must proceed material creation, and is absolutely non-material in its original form, in effect originated from a higher invisible pre-dimensional reality which preceded the Big Bang and the material universe, then this divine 'likeness' must refer to the invisible pre-materialised being of the creative source.

Furthermore, if we consider the principles of the macro and microcosm 'As above, so below' - it is far more likely that this statement refers to the divine nature of God (the creative source) which is present in all things, having created them from higher dimensions of reality.

If we can assume that this 'God' pre-existed the

universe, being the causative factor of it and all things in it, we must consider the fact that this God is non-physical and non-material, being beyond form, and of an altogether different nature. If then, we are 'made in his image', it is, I believe, the non-physical, invisible and divine aspect of God to which this statement refers. Exactly what this mystical and divine energy may be however, we shall explore in the next chapter.

Being essentially non-material by nature, furthermore, I doubt whether this 'God' has a single gender of maleness alone, which has been so popularised by the male dominated monotheistic religions, and which supplanted the primal and pagan religions where 'God' and 'Goddess' worship was not uncommon. Indeed, the Pontiff himself has now declared that God is possessed of both Male and Female energies!

Already intrinsic in Taoism, Buddhism and Confucianism, the concept of male and female energies - the polarity of Yin and Yang - is known as dualistic monism (the two originating from the one) which ultimately gives rise to the numerological concept of the Trinity - (1+2=3), which is also referred to in many religious texts, including those of Judaism, Christianity and Hinduism.

Now, although matter is finite, in the sense that all material forms are subject to birth, life and death, 'energy' itself can neither be created nor destroyed. It was never born, and it can never die. It just is. It precedes all forms, being the eternal creative source of matter, and in this context it can be considered infinite in its absolute sense. (I use the term 'energy' here to denote a 'higher' or causative 'divine' energy.)

Furthermore, this 'energy', according to Panpsychism, is present in every living thing and that all things have 'mind' or consciousness which is viewed as a fundamental feature of the Universe. Panpsychism postulates that consciousness did not evolve when our brains became sufficiently complex.

Instead it states that consciousness is inherent throughout the Cosmos. No living thing, animate or inanimate, can exist without this energy.

However, infinite or not, let us make no mistake, this 'energy' is not merely a by-product of creation, it is intrinsic to creation itself. Equally, energy also appears to have some form of intelligence or awareness, as well as the ability to organise matter into a multitude of forms and species. This is obvious and apparent if you look at Nature.

What quality is it in a moth, for instance, which has the intelligence to make that moth look identical to the tree on which it lives? Was it the intelligence of the insect itself that determined that it should look like its host tree, or did the very atoms and cells which themselves must be party to this intelligence, which determined their evolution?

However, 'Energy' is possessed of many qualities, and one of the qualities of universal energy that we need to investigate at this point is the nature and activity of Light. Although light was once believed to be corpuscular, that is, consisting of particles, it was later proved by the phenomenon of diffraction - which shows that the boundaries of shadows are diffused - to consist of waves. Although apparently contradictory in scientific terms, light appears to consist of both particles and waves both alternately and simultaneously.

Light itself may be regarded as a form of electromagnetic radiation consisting of oscillations of an electric and magnetic field. Light forms a narrow section of the electromagnetic spectrum and is perceived by normal vision on a wavelength range between approximately 390 nanometers (violet) to 740 nanometers (red). Substance and light are basically the same electro-magnetic energy and can be described as fields of force whose state is discernible as wave phenomena.

Light travels out from source at 186,000 miles per second and is, (apart from thought as we shall see later), the fastest moving force in the universe. The relative speed of light furthermore, has much to do with how we perceive the visible universe. Indeed, without light, we would perceive absolutely nothing! It is light that makes all things visible and manifest.

Without light, our world would be in darkness and no life would exist at all, being entirely dependent on it for biological function and creation. Light, so far as we are concerned, is synonymous with life itself, and without it, we would simply not exist. Everything we are and perceive is a direct result of the activity of light.

Now, as we are all no doubt aware, when we look into the night sky at the furthest most distant galaxies, we are looking at the universe as it was some millions or even billions of years ago. This is quite simply because light takes time to travel from one place to another, and the light which we perceive from these distant galaxies was actually emitted millions, and in some cases billions of years ago.

Much closer to home is our own Sun. The light from this, the closest star to us, takes roughly eight and a half minutes to reach us. Therefore, if the Sun were to go out now, at this very second, we would not be aware of it for another eight and a half minutes. Due to the nature and speed of light, therefore, there may be delays in between the time that an event actually occurs and the time you perceive the event happening. Furthermore, light and time can both be diffracted and distorted.

Albert Einstein's theory of Relativity demonstrates to us that we experience the phenomena of Time/Space, relative to the speed of light and the relative position of the observer. Einstein also discovered that the faster one moved relative to the speed of light, the slower time went.

This was demonstrated amply in his mathematical experiments which indicated that if a traveller was to embark on a journey into space moving faster and faster at a relative velocity, he would return to Earth to find that compared to, say, his ten relative years in space, Earth time had moved on at a relative ratio to his own, and had aged several hundred years.

Let us just assume for a moment that we could reach the speed of light ourselves. Hypothetically - remembering that time moves slower as the speed of light is approached - at the speed of light itself, time and space would simply cease to exist. We would be beyond Time/Space, and definitely nowhere within the visible universe!

Logically and conversely, if we reverse the process travelling slower and slower away from the Light, we would no doubt arrive in some strange place which the inhabitants called 'the material universe', having passed through several dimensions to get there.

Understanding light has a lot to do with understanding the manifested Universe and 'light' has a mystical, as well as physical connotation. The subject of 'Light' also appears over thirty times in The Bible, and the tenets of Christianity as defined by Emperor Constantine at the Council of Nicea in the 3rd Century AD contain these profound words in their texts; 'Before all Worlds. God with God. Light with Light'.

The secrets of creation are bound, symbolically, encoded in our ancient texts and religions, which at their very essence contain mystical knowledge about universal energy and creation which compare with, and sometimes exceed current scientific knowledge.

We also speak of people 'seeing the Light', or becoming 'Enlightened' to indicate a state of spiritual awareness or illumination. 'Going towards the Light' has also become a frequent phenomenon for those who have died and been brought back to life by modern medical science!

Nowhere is the mystical activity of light better conveyed, however, than in the Kabbalah - the mystical tradition which forms the basis of both Judaism and Christianity. The Kabbalah, which literally translated means 'to receive' or 'to reveal', has both a written and an oral tradition. The first four books of the Kabbalah appear as the first four chapters of the Old Testament.

The three main written texts of the Kabbalah are The Sepher Yetzirah, which describes God as indescribable, The Zohar, which is commonly referred to as the Book of Splendour, and which describes the Universe as an interconnected mass of particles governed by a higher force, and the Sepher Bahir - which is known as the Book of Brilliance - which describes the Universe as a multi-layered reality in which all parts are connected and where each and every part is governed by a higher law.

The teachings of the Kabbalah are visually encompassed in the Kabbalistic 'Tree of Life', a schematic diagram which represents the underlying blueprint and numerology of creation, and which best describes how God (Energy, Spirit, Light) manifests the physical Universe (matter) through a series of dimensions called Sephiroth. Simply put, 'God' which can at its essence be viewed as pure Energy, Consciousness or Light, must be slowed down through this series of transformers, (dimensions), to its slowest vibration which will then be perceived as matter, travelling through at least ten dimensions to get there.

Each of the ten Sephiroth also represent the numerological principles of creation from one to ten, and are equated with the ten major planetary influences which shape life on Earth. It is through the planetary energies that the ten main qualities of the Tree of Life manifest themselves.

The Tree of Life is then further subdivided vertically into three - by the Pillars of Judgment, Mildness and Mercy - representing the Trinity (Spirit, Mind, Body) and is subdivided again into Four Worlds which represent the four elements, Fire, Water, Earth and Air. Divided horizontally there are seven levels, which can be seen to relate to the seven chakras.

To the Kabbalistic or Occult numerologist, numbers

represent the primal organising principle that gives structure to the material universe, the seasons, the movements of the planets, and even harmony in music are all determined by numerical law. There is a rhythm to life, and these rhythms can in turn be measured as cycles, waves or vibrations, all of which are measurable by number.

As Pythagoras once observed, 'All things are number.'

Numbers and numerical law are at the very heart of creation. Numbers represent the ability of the One to self-divide and multiply in the same way that an embryo becomes a fully formed entity. It is through this self same process that the entire universe and all things that exist come into being.

Before Creation.

'Our birth is but a sleep and a forgetting.
The soul that rises with us, our life's star,
Hath had elsewhere it's setting,
And cometh from afar'.
Wordsworth. 'Ode to Immortality'.

Before you came kicking and screaming into this world, and made yourself visible and manifest, you spent nine months in your mother's womb, invisible and as yet unformed. Although you were unseen and virtually unknowable up until the time of your birth, there was no doubt in your mother's mind that you were in the process of creation.

This simple analogy might best describe the birth of the Universe, for, up until the point that it burst through the Ethers into the realms of the visible, it was in a state of gestation and formation, let us just say for now, in the womb of creation.

Now, it is quite apparent, if we follow the principles of the 'as above, so below' maxim relating to the relationship of the macrocosm and microcosm, that the universe itself must have been going through a similar process of conception, development and gestation before its own birth.

Somewhere, unseen and invisible to the naked eye, on an altogether different dimension, the universe was in a state of pre-birth up to the point of the Big Bang, which we will liken to the moment of birth, and the exact point at which the Universe exited the womb of the Universal Mother.

However, whatever we are, or were in the beginning, be it 'God', spirit, energy, atomic matter or otherwise, no one part of the Universe can be separated out from any of the other and at the point immediately prior to the Big Bang, all matter was contained in a state of unification, or oneness.

Previous to the visible and physical manifestation of the Universe which occurred at the Big Bang, however, the laws of physics as we know them break down and, as we are dealing with higher dimensions of reality, simply cease to apply. We are, in essence, now dealing with non-material principles that do not, and cannot conform to any known laws that apply in the physical dimension. Another set of criteria for understanding pre-creation of Universal energies is therefore required in order to discover exactly how or where matter might have existed before the Big Bang itself.

Of course, we know that matter did not originate in our own universe, because our own universe did not exist itself in physical form before the Big Bang. So where was all that universal substance before the Big Bang?

One way of explaining how and where the universal substance could have been or existed before the Big Bang is to consider the concept of higher dimensions - another state of existence in, or on which this universal energy or substance could actually pre-exist. Now, although the idea of higher dimensions may sound at first rather like science fiction, the concept of a multi-dimensional universe is now fast becoming science fact!

The first theory of higher dimensions was called the

Kaluza-Klein theory in which light, which can travel through the vacuum of space - unlike sound - was explained as vibrations in the fifth dimension. This was later advanced by physicists Michael Green and John Schwarz, who proved the consistency of this theory with the 'Superstring Theory', which suggested that all matter consisted of tiny vibrating strings.

Although it was once thought that atoms were indivisible fundamental particles, the discovery of neutrons, protons and electrons within the atom itself has now shown that atoms are not indivisible at all! Furthermore, scientists have recently discovered that neutrons, protons and electrons are not indivisible fundamental particles either, and are now known to contain three sets of even smaller particles that are known as Quarks.

Quarks themselves come in two varieties - briefly named the 'up' quark and the 'down' quark. A proton consists of two up-quarks and a down-quark and a neutron consists of two down-quarks and an-up quark! Furthermore, everything in the universe appears to be made from combinations of electrons, and up-and-down quarks.

However, that is not the end of the story as far as sub-dividing atoms and fundamental particles is concerned, and in the 1950's Frederick Reins and Clyde Cowan found conclusive experimental evidence for a fourth kind of fundamental particle called a neutrino - billions of which are ejected into space by the sun, and are passing through us all at this very moment!

Scientists have now also discovered four more quarks, known as 'charm', 'strange', 'bottom' and 'top', and a heavy electron called a 'tau', as well as two other particles called muon neutrinos and tau neutrinos. According to Superstring Theory, if we could examine these particles with even greater precision we would find that they all ultimately consisted of tiny vibrating loops!

In his book 'Hyperspace', Michio Kaku, Professor of

Theoretical Physics at New York University writes; 'The deeper we probe into the nature of subatomic particles, the more particles we find. The current 'zoo' of subatomic particles numbers several hundred and their properties fill entire volumes. Even with the Standard Model, we are left with a bewildering number of elementary particles.

String theory answers this question because the string, about 100 billion billion times smaller than a proton, is vibrating; each mode of vibration represents a distinct resonance or particle. The string is so incredibly tiny that, from a distance, the resonance of a string and a particle are indistinguishable. Only when we somehow magnify the particle can we see that it is not a point at all, but a mode of a vibrating string. In this picture, each subatomic particle corresponds to a distinct resonance that vibrates only at a distinct frequency.'

Although we shall be referring to Superstring Theory again in this book - suffice it to say that according to quantum physicists, in order for this theory to work, there must be further dimensions to explain and account for these quantum phenomena. The precise number of dimensions predicted? Ten!

Although Superstring Theory represents the latest developments in quantum physics, the concept of a ten-dimensional universe already exists, as we have seen, within Kabbalistic tradition and teaching.

In the 'Hall of the Gods', referring to esoteric numerology, author Nigel Appleby writes; 'In the Jewish mystical tradition of the Kabbalah, it is taught that God has 10 faces and that followers of the faith should know them all. In quantum science, where scientists study atomic particles in an accelerator vacuum, (a vacuum is basically 'nothing' which consists of particles that spontaneously create and annihilate themselves), some authorities have given 37 as the suspected number of particle fields. In Fadic terms, this number is 3+7, which is equal to 10, i.e. in Pythagorean reasoning everything (1) and nothing (0). So according to both a specific religious interpretation and the tenets of scientific quantum theory, there was 'everything' residing in 'nothing' - symbolised by the number 10. From the very start of the universe itself, the basic building block number was to be 10.'

The concept of a ten-dimensional universe is a theory now gaining credibility in the scientific world, and in his book 'The Tenth Dimension' Japanese physicist Mikio Kaku describes the theory of a ten dimensional Universe, and his diagram of how he perceives these multi-dimensional universes is remarkably similar to the Tree of Life of the Kabbalah.

In his own search for a Unified Field Theory, a grand theory that would unite all theories into one, renowned physicist Professor Paul Davis, in his book 'Superforce' explains how the bewildering array of subatomic particles form abstract patterns and mathematical symmetries, suggesting deep linkages. He explains how the forces that act between these particles may require the existence of unseen extra dimensions of space and how these particles may be linked by an invisible 'Superforce' which permeates all of Creation. Could this 'Superforce' be God?

In 'The Physics of Immortality', author Frank Tipler presents a purely scientific and mathematical argument for the existence of God and an afterlife. In his opening pages he writes: 'When I began my career as a cosmologist I was a convinced atheist. I never in my wildest dreams imagined that one day I would be writing a book purporting to show that the central claims of Judeo-Christian theology are in fact true, that these claims are straightforward deductions of the laws of physics as we now understand them. I have been forced into these conclusions by the inexorable logic of my own special branch of physics'.

Although the superstring theory represents the latest theoretical developments in quantum physics, and may even have its own opponents from within the scientific camp itself, (string theory remains a theory because we do not have the technology that can magnify the relevant subatomic particles 100 billion, billion times), I believe that ancient mystical systems will wholeheartedly welcome Superstring Theory into their camp, as the Kabbalah and its concepts of integrated subatomic systems and ten dimensional universes has already shown.

The following passage from the Hindu scriptures however - the Baghavad Gita - may suggest that the concept of superstring theory was already incorporated into Sanskrit texts thousands of years beforehand. In Baghavad Gita, Brahma (God) manifesting as Lord Krishna explains; 'I am Arjuna, the highest principle of transcendence, and there is nothing greater than me. Everything that be, rests on my energies, exactly like pearls on a thread'. (BG 7.7)

In the Beginning...

'In the beginning was the word, as the word was with God and the word was God'.
The Gospel of St. John (1:1)

I do not, for one moment, hope or suppose I am able to describe to you in any meaningful way the true nature and being of 'God' or exactly how this God came to manifest the visible Universe. The Kabbalah, as well as many other great religious and mystical systems, philosophies and texts, remind us that God is indescribable, and that if you can describe God, then it is not God at all.

I must be honest and admit, therefore, that all I can hope do in these pages is to allude to the nature and being of God, the initial and the indescribable source of all Life, since adjectives and verbs are obviously insufficient to penetrate the nature and being of the Creator, who is beyond form and certainly beyond the reach of the human intellect.

As we shall see later however, although the human intellect is incapable of understanding God, God can be known through the internalised mystical experience of vision, revelation and the transcendental experience. Words, however, are often insufficient to describe the personal mystical revelation, as it consists of experiences for which we have no suitable language.

Before we can address the inward mystical understanding of God, however, there is one more aspect of creation that we need to investigate, and that is the nature of Sound. Sound is essentially a vibration that is carried through a medium such as air or water (sound cannot exist in a vacuum) and is experienced by the human ear on a range between 20 and 20,000 hertz (audio frequencies).

According to the New Testament, sound was at the heart of creation and we are told 'In the beginning was the Word and the Word was with God, and the Word was God' (John 1:1). According to ancient sacred Hindu literature Aum (Om) is said to have been the word, sound or vibration that God uttered to bring forth the materialised Universe.

What effect does sound have on matter?

The eighteenth century German physicist Ernest Chiadni first observed the effects of sound when he scattered sand on steel discs and observed the changing patterns produced when various notes were played on a violin.

Earlier this decade and inspired by Chiadni's experiments, Dr Hans Jenny began working with a 'Tonoscope', that is, a machine which transforms sounds into visual images on a video screen. When Dr Jenny recited the sacred Hindu word Aum (Om) through the Tonoscope, he was amazed to find that the pattern that emerged was a perfect circle filled with concentric triangles and squares.

This pattern coincided with the frequency pattern of diminishing harmonics, and was identical to the Buddhist 'Shri Yantra', a mandala that is said to represent the moment of creation, which was said to have begun with the utterance of the sacred sound 'Om'. Further experimentation by Dr. Jenny also showed that different notes were able to create different shapes, many of them intrinsic and essential in creation, such as spirals and octagons. He also found the last note of Handel's Messiah produced a perfectly formed six pointed star!

It would appear then that at the subatomic level, atoms closely resemble oscillators that respond to different frequencies. According to physicist Donald Hatch, 'we are now discovering that all matter is in a state of vibration and that the Universe is composed not of matter, but of music!'

Now, back to superstring theory and our tiny vibrating loops. Could it possibly be that these vibrating loops are making musical notes with their vibrations that actually organise matter into cohesive and mathematically correct living art forms! Are these tiny superstrings emitting tones that materialise form into different shapes and patterns?

In his book 'The Music of Life', Hazrat Inayat Khan, a Sufi mystic writes; 'A keen observation shows that the whole universe is a single mechanism working by the law of rhythm. The law of rhythm is a great law, which is hidden behind nature. It is in accordance with this law that every form is made and that every condition manifests to view. The creation therefore is not merely a phenomenon of vibrations without any restrictions. If there were no rhythm, if it were not for the law of rhythm we would not have distinct forms and intelligible conditions. There is no movement which has no sound and there is no sound which has no rhythm.'

But where does this sound and rhythm come from? It is coming from within the vibrating 'superstrings' that exist within every one of the atoms that makes up our totality! Within each of us at this moment, if we could but perceive, is an incredible symphony being played, each of our differing mathematical proportions - of this or that atomic structure - producing the sum total of ourselves. Each of us having our own individual vibration, our own unique orchestra and sound, and our own unique composition to play!

In his book 'The Elegant Universe', Brian Greene, Professor of Physics and Mathematics at both Columbia and Cornell universities writes; 'Music has long since provided the metaphors of choice for those puzzling over the questions of cosmic concern. From the ancient Pythagorean 'music of the spheres' to the 'harmonies of nature' that have guided inquiry through the ages, we have collectively sought the song of nature in the gentle wanderings of celestial bodies and the riotous fulminations of subatomic particles.

With the discovery of Superstring Theory, musical metaphors take on a startling reality, for the theory suggests that the microscopic landscape is suffused with tiny strings whose vibrational patterns orchestrate the evolution of the cosmos. The winds of change, according to superstring theory, gust through an Aeolian universe'.

However, not even the most sophisticated of instruments can play itself! It cannot utter a note unless there is a player or a plucker, a vibration will simply not occur in the instrument itself until it has been agitated by some outside force.

So what is causing our tiny little superstrings to vibrate? Let us go back to the reference from the Bible where we are told: 'In the beginning was the word and the word was with God and the word was God' - what could this possibly mean? As we have already seen, according to Hindu philosophy and teachings, God simply uttered the sacred sound of 'Om' and brought the whole material universe into being! Was this the primary sound that set all things into vibrating motion?

Light and sound are both essentially disturbances on the electromagnetic spectrum in a given medium, and depend on two things for their existence, a source which has activated the vibrations in the first place, and a source which can act as a receiver for those vibrations.

It would appear then, that although both light and sound are primal energy forms in their own right they would both still depend on an initialising source to stimulate the initial vibration that caused either one of them to exist in the first place. What was this initialising factor?

If we go back to maxim, as above, so below, there is one other energy that we should consider if we are to understand something about creation and that is, the power of thought. Thought, in its essence, must be viewed as a power in its own right. Thought - the power of creative visualisation - is behind every work of literature, art, music, invention and creation. Thought, in itself, is an illusive power that defies definition.

Think about it. As you sit and read this book, realise that each word that is written on this paper is being read out loud to yourself in your mind. You are thinking the words out loud in your head. No one else can hear them but you. Think about where those thoughts are coming from. Somewhere in your being is the source of the thinker, and although we tend to associate thought solely with brain activity, this, as we shall see later, is simply not the case.

Thought is a vital prerequisite for creativity. Nothing that is ever done, not the writing or reading of this book, or any other thing you care to imagine was ever done without an initialising thought. Whatever you or I do is always preceded by a thought. Quite simply, we think about something, and then we do it. 'I think I'll do this', or 'I think I'll do that'. We create it in our minds, as a concept, before materialising the thought into deed.

Thought and the power of visualisation and creation are closely linked and in the simple, but profound words of the l6th Century French philosopher Descartes, 'I think, therefore I am.'

Thought is the creative source of the self. Thought is pure creative power. Nothing is ever done without it. Without thought, creative and conscious, there is no motion, no creation. Thought, as we shall see later, travels faster than light. Thought can be instantaneously received regardless of the distance of the thinker and the receiver. Thought precedes all things. Thought, as a form of Energy, creates ideas that are translated into action and then materialised as form.

Perhaps then, it is the same with the creative source of the whole Universe, perhaps the 'Mind of God' conceived of a multi-layered universe, with planets and stars, mountains and seas, sweet smelling orchids and a variety of living things.

In his book 'Christ and the Cosmos' Professor E.H. Andrews writes: 'God is the common source and originator of all things. He did not take some pre-existing substance and transform it into the world. This is the whole thrust of the doctrine of 'ex nihilo' creation. He created this entire universe from pure spirit, from pure thought, from pure mind'.

The Creative Source thought about not being alone in the Void, and created a seed thought within itself which emitted a vibration which in turn emitted sound and light, and brought the whole of the material Universe into being.

The Big Bang.

'I cannot believe God plays dice with the Universe'.
Albert Einstein.

Approximately 14 billion years ago our Universe was born. Brought forth from the invisible dimensions that preceded material creation, the darkness and silence that must have existed at that moment echoed to an incomprehensible explosion, which in turn spewed matter forth into an expanding space.

One cannot even begin to imagine what it must have been like, or what could have existed in the emptiness before that time. Our intellects can only comprehend that which, like our physical selves, has a beginning, a middle and an end. It is virtually impossible, therefore, by any normal processes, to understand the possibility or reality of an existence without beginning or end, in other words, infinite and eternal.

In the sacred and mystical texts of Hinduism, the Big Bang is known as the Divine Breath, and it is said that Brahma (God) breathes in and out an endless cycle of Universes for all eternity. According to these ancient Sanskrit texts, the Universe is created many times and destroyed equally in an endless cycle.

The birth of our Universe is but one incarnation of many such Universes which existed aeons ago in the past, and which will exist again aeons into the future. These intervals may be beyond our comprehension, the interval between the expansion and contraction of one Universe maybe taking trillions of years to complete, exploding and imploding in regular cycles.

Who is to say how many times this has already occurred? Perhaps energy and matter are eternally co-joined, as in the Chinese symbol of Tai Chi, the Yin and Yang, bound together, inseparably, symbolising an endless cycle of creation and destruction which continues for all eternity. Spirit and matter endlessly co-joined in a cosmic dance!

Perhaps it is part and parcel of the nature of energy to create, and following the cycles of creation and destruction, the universe returns impeccably to it's source, to be spewed out again and again, for all eternity.

At the point immediately prior to the Big Bang, however, which is the beginning of the Universe as we know it, all matter was contained in a singularity, that is, a point where all the laws of physics as we know them break down.

Having been in the process of creation since the Divine Thought conceived of it, the One, having undergone the many mystical and alchemical processes required to materialise pure thought into form, was ready to give birth, in this case to an entire Universe.

The One was about to become the Many. By a process of self-division and multiplication, as in the cellular division of an embryonic form, the original One had self-multiplied itself and was ready to manifest onto the material plane.

One can only assume that matter burst through the Ethers, perhaps through something similar to a White Hole - the opposite of a black hole which sucks matter in - after completing its gestation period of dimensional formation, finally ejecting itself out from the womb of creation.

At the precise moment the Big Bang occurred, immense amounts of matter began expanding into an evolving space, and slowly over billions of years, this cosmic matter, which consisted of very fine dust, began to condense into Stars, Planets, and, eventually all life forms as we know them, including ourselves.

Once the Universe had condensed sufficiently, it formed billions of galaxies and stars, including the spiral galaxy in which we now live, and our individual Solar System, that is, The Sun and nine known Planets, Mercury Venus, Earth (Moon) Mars, Jupiter, Saturn, Neptune, Uranus and Pluto*.

And the rest, as they say, is history.

(*Although Pluto is no longer considered a planet and was demoted in part thanks to one Neil DeGrasse Tyson, to astrologers, Pluto still is, and always will be, very much a fully fledged planet!)

Astrology.

'Had we never seen the stars, and the sun, and the heavens, none of the words which we have spoken about the Universe would ever have been uttered. But now the sight of day and night, and the months and revolutions of the years, have created number, and have given us a conception of Time; and the power of enquiring about the nature of the Universe; and from this source we have derived Philosophy, than which no greater good was, or will be given by the gods to mortal Man.'

Plato: 'The Timaeus'.

Astrology has evolved from a time immemorial as a sacred mathematical and metaphysical science revolving around the study and interpretation of universal cycles and rhythms as seen from our Earth bound perspective.

    We depend entirely on the Sun, Moon and Planets for life as we know it, and their precisely predictable cycles are the very cause of everything we know and hold dear. Spring, summer, autumn, winter, night and day, the rise and fall of the tides, all are the result of precise and predictable mathematical variations in planetary movements, including the rotation of the Earth on its own axis.

    The vast planetary bodies are continually affecting our personal and planetary biospheres as they radiate their enormous electro-magnetic energies into space. Quantum physics tells us that even one microscopic electron on the far side of the universe can affect everything in the universe simultaneously. How much greater then will the effect of the planetary movements be - which are vaster by far?

    Any movement or expression of energy either internally within the atom, or externally in outer space - will have an instant reaction and affect everything, everywhere within the universe simultaneously. Earth is not exempt from this equation and we are being continually bombarded from outer space by cosmic forces to which the Earth's magnetosphere is extremely sensitive. These cosmic forces are passed on through the magnetosphere of the Earth to all individual life forms that (as we shall see in the next chapter) also possess their own unique electro-magnetic force field.

    Furthermore, according to String Theory, all matter in the universe - including the planetary bodies - is composed of microscopic atoms that in turn contain infinitesimally minute vibrating strings! The planetary bodies themselves are so vast, and must contain so many atoms, each with its own tiny vibrating superstring, that they must be making quite a silent noise out there in space! Like everything else, the planets are in a state of vibration.

The Planets, which are essentially vast orbs of singing, vibrating atomic matter - as we shall see shortly - are rotating at different orbital velocities and relative distances from a communal sun, which itself is a giant humming thermo-nuclear reactor which is continually creating fundamental articles which are the very breakfast of life!

We on Earth, the third rock from the Sun, are orbiting this great giant star at approximately 65,000 mph., accompanied by a relatively small lunar satellite that we call The Moon. The Sun, in its turn, is just one of billions of similar suns that are all rotating around the centre of our own spiral galaxy, which we call simply 'The Milky Way'. The Milky Way in its turn is spiralling onwards through space accompanied by a billion other visible galaxies. (Because the universe is expanding, however, all these stars and galaxies will one day entirely disappear from view, and if any life still exists on Earth billions of years hence, they will not share the joy of the star-filled night sky that we currently take for granted.)

For the purposes of Astrology however, we confine ourselves to understanding the planetary cycles within our own Solar System, from an Earthly perspective.Whereas it is the job of Astronomy to define planets in terms of their the physical properties, weight, mass, rotational and orbital velocities, chemical composition and so on, it is the job of Astrology to define the effects of these planetary bodies and their movements on human destiny, both individual and collective.

According to the basic principles of Astrology, each planet has its own character, or 'tone', which affects life on Earth and is said to 'govern' a particular area of life. In mythology, the planetary energies were often personified via a series of 'Gods', thus evolved the characters of Venus -The Goddess of Love, Mars - The God of War and so on.

These archetypes are found in many global religious mythologies, and although these personifications take on the characteristics of their own cultures, they still retain the essence of the planetary personalities.

Far from being inert spheres in space, however, all the planets including our Earth have vast networks of electrical currents running in their deepest interiors. On Earth, these energy lines are sometimes known as ley-lines and the most well known of these magnetic currents runs through Stonehenge, Glastonbury and St. Michael's Mount in the west counties of England.

For the purposes of terrestrial astrology, however, we are primarily concerned with the effects that these energies have on life on Earth as we are magnetically influenced, enhanced or disturbed by their immense energies as they orbit the Sun.

These magnetic variations, according to astrologers, directly influence life on Earth and are a medium for the direct expression of an organised universal will, consciousness or mind - God in other words.

In a sense, it is through the planets that we can pre-determine cosmic activity, the planets can be seen to be acting as a sort of 'pre-set computer code' for creation. It is through astrology that we can understand some of the secrets of this code. The signs the planets are in, the angles they make to each other and the house system, all determine the arrangement and interaction of electro-magnetic forces that we receive on Earth.

Before we take a closer look at Astrology, however, let us firstly take a simple analogy to help us imagine what these invisible magnetic lines of force in space might actually look like as the planets transit the heavens.

To do this, firstly imagine a calm pond with a smooth surface. When you have done that, toss in a pebble. Now - watch the ripples as they expand out from the point of impact, as the pebble travels through the medium of the water! This is how it might look if we could actually see the invisible lines of force coming from the planets, as they disturb the medium of space, in the same way that the pebbles disturbed the water! The size, shape, speed and nature of the individual planet will have a different effect on the shape of the disturbance in the medium through which it is travelling.

To continue our analogy, let us now throw ten pebbles into the pond at the same time at the relevant positions - to represent the Sun, Earth and all the planets - and imagine how the ripples will look on the surface of the pond now! Can you see how the vibrations from the ten impacts overlap and interact on the surface making lovely patterns!

Lastly, imagine these pebbles could assume an orbit around a central pebble and be thrown into the water together - can you see all the different patterns the pebbles (planets) make as they follow their spiral trajectory through the water (space)? Remember, we all live on the 3rd pebble - watch how the vibrations affect us!

The idea that the heavenly bodies can affect life on Earth is the primary principle of Astrology, calculating the planetary movements is the mathematics and geometry of it, and interpreting the effects of those movements is the science and art of it.

The practice and belief in Astrology goes back to the most ancient of days when our first ancestors began to study the heavens and found patterns and cycles that could be precisely predicted and which ultimately became the very foundation of timekeeping as we know it. Years, months, and days are all measured and created by planetary cycles.

In early Britain, it was the job of the Druids to keep abreast of these cycles and keep accurate time, which was essential for the organisation of society, and which they did with the assistance of Stonehenge. Astrology also became a science of prophecy, from the prediction of eclipses, equinoxes and solstices, to the prediction of personal and collective events through the interpretation of the signs in the heavens.

Working on the maxim of 'As above, so Below', Earth

was seen as a mirror of the cosmic world. The phrase 'As above, So below' comes from the beginning of the Emerald Table and describes an entire system of arcane knowledge that was inscribed upon the tablet in cryptic wording by Hermes Trismegistus. The significance of this phrase is that it is believed to hold the key to all mysteries.

Trismegistus equated God to the Universe, the Universe to Man and Man to the atom. In other words, there is no separation between God and Man as God is omnipresent in all things. Nothing, literally nothing exists without divine energy and that divine energy, so far as Earth is concerned, forms the blueprint of all that is to follow. Astrology holds the key to the individual blueprint.

'It's a load of Twaddle!' Thus read the headline of a newspaper article on astrology that appeared in the London Daily Telegraph, written by science editor Dr Roger Highfield. 'Pure Bunkum!' 'Codswallop!' 'Superstitious mumbo-jumbo!' These are a variety of uneducated opinions on Astrology including one from the late lovable pope of British Astronomy, Patrick Moore.

Once wedded in the same marriage bed, Astronomy and Astrology are still suffering the effects of a cultural divorce which began with the separation of the sacred sciences - alchemy, magic, metaphysics, mathematics, geometry, astronomy and astrology - which was due in part to the birth of modern materialistic science and in part to the accompanying collective shift in social values from spiritual to material.

At the same time, our materialism and technology was taking us further and further away from the countryside, our traditions, our knowledge of the land, the night sky, the rising of the sun and moon - away from nature - and further and further into the barren drabness of our cities.

Professor Adrian Furnham of University College London says that 'People believe in Astrology because they fall victim to the fallacy of personal validation'. (Does that mean

I'm not a Sagittarian, I only think I'm one?)

Professor Richard Dawkin who says 'Let us not go back to a dark age of superstition and unreason' - is also equally agitated by the rising popularity of Astrology, and he has made precisely the same accusations against religion, which he dismisses as thoroughly as he does Astrology.

Equally sceptical Dr. Susan Blackmore, Senior Lecturer in Psychology at the University of West England believes that; 'the human mind is made to make connections, and people will look for them in everything. This is what makes us clever and able to understand the world, and this is what trips us up when we read significance into horoscopes'.

Dr. Blackmore believes that people who turn to astrology tend to be those with 'least control over their own livesí' such as, and I quote 'women who read women's magazines'! Susan Blackmore also believes that 'Astrologers trade on the mind's attempts to understand the world that have been honed throughout evolution'.

When confronted with similarly uneducated critics, Sir Isaac Newton famously and politely replied 'I Sir have studied the subject, you have not.'

Now, if you are not altogether familiar with the principles of Astrology, and you too feel an instant need to doubt the ability of the planets to affect personality, destiny and life on Earth, just consider this for a moment before we go any further - what do you think would happen to you, your family, your pet dog, and the rest of life on Earth if the Sun were to go out right now?

Exactly. We'd all be dead in no time. We would all die. I can comfortably predict that. We depend on the Sun for life itself. Our very ecology and biology is as linked and dependent on the Sun for the food of life, as a child is to its mother for sustenance.

Now, our lunar satellite - what would happen if the Moon were to suddenly escape from the Earth's gravity and disappear into Space? If you don't know, I'll tell you! The

Oceans would swell into gigantic tidal waves for one last time - before they finally settled into one giant stagnant unmoving pond! We'd probably all be dead in no time. We would all die. (The Moon is actually moving out of Earth's orbit at a slow but steady pace. This will mean nothing to us now, but in millions of years time the Moon will actually finally leave Earth's orbit forever, with potentially devastating consequences for all life forms on Earth.)

Although the edifices of Stonehenge and the Pyramids remain as a lasting testament to a time in our pre-history when ancient civilizations displayed an arcane knowledge of the metaphysical sciences which far exceeds our own, the science and art of Astrology has found no grace and favour in our modern Western society until relatively recent in our history.

Widely practised in Britain until the reign of Queen Elizabeth I of England, who employed her own personal Astrologer, one Dr. John Dee, and his assistant, Irish mystic Edward Kelly, to help guide her affairs of State, Astrology finally fell out of favour shortly after her death, in l603.

Far from being a 'weak and aimless woman' who had little or no meaning and control over her own life - as Dr Blackmore has suggested - Elizabeth I was a competent head of state for over forty years and one of the most successful monarchs in British History.

Once voted 'Woman of the Millennium', was Elizabeth's success due in part due to the wise council of her astrologer and mystic who often warned her of impending danger, including the subversive Catholic plot to kill her, and place her cousin Mary Queen of Scots on the throne of England? If we can take Elizabeth I, as a case in point, I think Ms Blackmore's statement is grossly unfounded.

Although a keen astrologer and alchemist himself, born shortly after the Tudor period came to an end, Sir Issac Newton later came to discover the physical laws that were to become the bedrock of modern materialistic science.

In the meantime, and having lost favour with the theoretical scientific academics of the post-industrial revolution who found no room for the mystical and metaphysical in their test tubes and theorems, the Church was entering an era of unprecedented brutality driven by a puritanical fervour. For over four hundred years, Britain, Europe and America entered a period of history known best for its obsession with 'witches' and 'witchcraft'. (NB: a 'witch' is the female equivalent of the male 'wizard' or magician.)

This period was characterised by a puritanical religious zeal orchestrated by the Church itself, whereby thousands of ordinary people, mainly women, were harassed, persecuted, tortured and murdered for practising any of the so-called heretic 'pagan' arts, including, of course, astrology. (NB. Women were seen as being more easily seduced into doing 'the devil's work' after Eve gave Adam the proverbial apple. He could have said 'no'!)

It was not until the Witchcraft Act was repealed in Britain in 1951 that Astrology - and Astrologers - could safely resurface from over four hundred years of political and religious oppression, having been discredited by both formal religious doctrine and scientific dogma.

Today however, with Astrology having been largely re-introduced to the West with the influx of Eastern religions and philosophies in the sixties, the study of the stars has once again captured the imagination of the collective psyche. Today literally millions of people all over the world read their 'Stars', in the daily papers and consult Astrologers regularly as a matter of course, hoping for some Heavenly guidance in their daily lives.

The words 'What sign are you?' has become the by-phrase of our times. In one word you can tell people a lot about yourself and your character 'I'm an Aries' (Loud and bossy!). 'I'm a Scorpio' (Sex maniac!), 'I'm a Virgo' (squeaky clean!), and so on. Astrology has become the human jungle way of 'sniffing' each other out now perhaps, that we have become so deodorised. Astrology has become very much the language of personality in today's world.

As well as describing character from the birth chart, (which is a map of the heavens at the time of individual birth), astrology is also helpful in predicting trends. Seeking some form of divine, honest and meaningful guidance in their daily lives, people are now turning to Astrology to interpret the meaning and mood of the Heavens and to fulfil their spiritual and emotional needs.

The increasing popularity of Astrology is, of course, much to the annoyance of the Church who say 'Instead of turning to the Church for guidance, we are turning to our horoscopes, taking solace in New Age philosophy', and that our quest for the 'quick fix, pick and mix' belief system is causing the very fabric of our society to fall apart. 'The whole of civilisation' says the Church, 'is on the verge of disintegration'.

In response to the accusations against Astrology, the late Jonathan Cainer, one of Britain's foremost newspaper astrologers has this to say; 'If the surge of interest in astrology, healing, homoeopathy, herbalism, self awareness or cosmic attunement is a sign of anything at all, it is that people are seeking greater depth and meaning in their lives. What should the Church have to fear from this?'

Although Astrology is dismissed by formal Christian teaching as a superstitious 'pagan' science which has no place in the 'real' worship of God - being both blasphemous and heretical - I am now more inclined to believe that the political and economic motivations and ambitions of the 'Church' which firstly arrived in Britain with a conquering Roman army, have led them away from the real meaning and purpose of Christ's teachings and message.

To support this claim, I would firstly like to refer you to the passage in the Old Testament of the Bible which appears in Genesis 1:14 and which reads; 'Then God said 'Let there be lights in the firmament of the heavens to divide the day from the night, and let them be for signs.' Now, what interpretation can you put on this? There can be no other! The planets and stars, the sun and moon, according to the Bible itself, are also for signs, or omens of the destiny on Earth to be read in the Heavens.

Indeed, the history of Christianity itself begins with one of these great heavenly astrological signs. The Three Wise Men - Astrologers - from the East, had found, interpreted and followed the New Star to Bethlehem to find the new-born babe. The Birth of Christ was predicted and clearly written in the Heavens. The Three Wise Men had found the signs clearly written in the stars.

In his missing years Christ is also reported to have travelled to India to study with the Hindu Brahmins. Christ's visit to India is recorded in the 'Pali' manuscript which was found in 1887 in Himis, the largest monastery in Ladakh just west of Tibet. He was also said to have travelled to Britain with his tin-mining uncle, Joseph of Arimathea, eventually arriving in Glastonbury to study the art of sacrifice (self) from the Druids. Blake's immortal hymn, Jerusalem, 'And did those feet in ancient times, Walk upon England's pastures green? And was the Holy Lamb of God on England's pleasant pastures seen?' is a reference to this myth.

Whether or not Christ travelled to India or Britain, he was in no way a stranger to the alchemical, esoteric and astrological knowledge and traditions of the time. By the time of Christ's birth, astrology was already well established as a metaphysical science. We cannot tell when the moment of awareness dawned, when our earliest ancestors first recorded the comings and goings of the heavenly bodies, but at some time in our distant past, humankind generated a vast knowledge of the heavenly movements and their relationships to events on Earth.

Long before the telescope was invented, the outer planets were well known to many ancient cultures even though they did not, so far as we know, have access to any sophisticated optical equipment. Nonetheless, early cultures show a remarkable understanding and knowledge that has sometimes proven superior to our own.

The Dogon tribe in Africa are amongst them, and have detailed star maps of Sirius which are absolutely invisible to the naked eye, which date back approximately 10,000 years. These, they say, were given to them by visitors from outer space. (Although it is not my job in this book to discuss the influence of Alien Intelligence on Earth culture, let us just say for now that many unanswered questions remain on this matter.)

To our earliest ancestors, the Sun was the obvious source of all life on Earth - giver of light, and warmth - without which, no living thing could be, grow or flourish. In many early societies the Sun became the focus of religious worship and many Solar temples were built in sun worshipping cultures. Was Stonehenge a Solar Temple? Why were the pyramids built on solar geometric principles? Did the ancients have vital knowledge about the universe and our solar system that we have lost?

In his book 'Sun of God', author Gregory Sams argues that the Sun is not just 'an accidental ball of gas' and that it has intelligence, will and consciousness. In the same way that the Gaia theory postulates the idea of a conscious earth, Sams makes us aware that the Sun has a living stellar consciousness with intrinsic purpose. Using evidence gathered from both solar science and ancient philosophies, he suggests that solar characteristics are 'indicative of an organised awareness'. Furthermore, given our physical bodies are made from the remnants of several previous Suns - it is surely no wonder that we still retain a subatomic connection to - not only our own sun - but all suns and stars and indeed to all living things through our common stellar origins.

Although our early forefathers may have known little of the material chemical composition of the Sun, they knew instinctively that it was essential to life. It is only recently that we have discovered that our Sun is in essence a vast thermo-nuclear reactor, which is continually spitting out fundamental particles into the atmosphere enabling creation and life on Earth to flourish.

Now, although our own Sun is just one of ten billion stars on the edge of a spiral galaxy, and our galaxy itself is but one of ten billion galaxies visible within the Universe, it is to us the very centre of the solar system which we call home, and is the centre around which all the planets, including our Earth, orbit.

The Sun doesn't just emit heat and light, however, it also emits other forms of electro-magnetic radiation - as well as emitting fundamental particles - and this outpouring of ionised gases and their associated magnetic field is called 'The Solar Wind'. The solar wind is a hot plasma, an electrically charged mixture of ions whose source is the Sun's corona. This highly charged gas flows outward during coronal mass ejections at approximately 500 kilometres per second to a distance four times greater than the orbital path of Pluto - which is known as the helio-pause. Because of the Sun's rotation, the emissions form into an Archimedean spiral.

When the Solar Wind hits the Earth's magnetosphere it distorts it with enormous force, depending on the speed and density of solar emissions. The varying amount of Solar emissions - magnetic plasma - which enter Earth's atmosphere will therefore have a direct effect on the magnetic fields of all other life forms, including us.

Medical research has also shown that people born at different phases of sun spot activity - either high or low - had different personality traits - optimists being born during low sunspot activity, and depressives and pessimists born during high sunspot activity.

According to recent evidence from NASA, when solar radiations are at their strongest, heart attacks (amongst other things) are most frequent. In Astrology, the Sun is said to rule the sign of Leo - which governs (amongst other things) the heart!

Symbolically, scientists have now detected what sounds like an actual heartbeat coming from our Sun, having observed a repeating pattern in a solar flare more than 3,000 miles above its surface. The beat repeated every 10 to 20 seconds. These pulses known as quasi periodic pulsations (QPP) are important for understanding how energy is released from the Sun during Solar Flare activity. No doubt, this Solar activity will be picked up by our own magnetic fields as they react to cosmic activity.

Astrologers have long since known the secrets of the planets and their relationship to life on Earth.

According to A.T.Mann in his Book 'The Round Art', 'The solar system can be considered a huge step down transformer of solar energy which produces an effective range of magnetic fields governing all life processes on Earth. The tracks of the planetary bodies around the Sun create magnetic fields charged with induced current and the Earth receives these as an archetypal magnet.'

Because of the electromagnetic nature of the planets and the Sun, (whose own surface is covered in dancing positive and negative magnetic loops), the magnetic ripples, or waves, of energy which are continually emitted from them cause electrical interference on the cosmic medium of space - in this case our solar system - which is directly felt on Earth as a form of wave energy. This wave energy penetrates the Earth's biosphere as electrical interference or agitation of the Earth's electrical field. This can be detected as radio interference.

In 1951, J.H. Nelson, an engineer with RCA was asked to investigate a problem RCA was having with its radio reception quality that seemed to vary with sun spot activity. After careful research, Nelson also discovered that the days when interference was at its worst were on days when any of the planets were forming specific angles of 0, 90, or 180 degrees to the Sun.

Since planetary movement is entirely predictable, based on the exact mathematical calculations of the orbital paths of the planets, this knowledge of future planetary activity and angular relationships enabled RCA to determine days on which radio reception would be affected - in advance!

In exactly the same way as Nelson was able to calculate future planetary movement and associated interference, by tracking the movements of the planets and observing their angular relationships to each other and the sun, Astrologers are also said to be able to chart future events, either individually or collectively.

By drawing a map of the heavens at the relevant point in time adept astrologers are also able to describe, not only an individual, but his or her life's journey in detail, by predicting the relationship between the positions of the planets at birth - each planet having its own unique magnetic force field, character and signature - and their continuing orbital relationships to each other in relation to the original birth chart.

In Astrology, each planet is said to have a different effect on the Earth's magnetosphere, and to govern one archetypal area of creation. The Moon for instance, directly affects us through its gravitational effect on water that produces the tides of the Oceans. Scientists have recently measured the magnetic pull of the moon in a cup of tea! How much more then, the pull on our bodies which are some two-thirds water? Does this explain the tendency towards 'Lunacy' at full moons, when the magnetic pull of the Moon on terrestrial liquid is the strongest, and the crime and birth rate is highest?

The Moon also has a direct effect on cycles of glandular activity in the body, in particular, the human female menstrual cycle which is 28 days long - identical in length to the Lunar orbit around Earth. Many aquatic species, including coral, respond entirely to lunar cycles to determine spawning and mating patterns, and even when displaced in laboratory experiments, their body clocks remain entirely in tune with lunar magnetism. In these experiments, simple life forms were able to distinguish when the Moon was full, and directly overhead in readiness for spawning and mating, even though their captor investigators had done their best to trick them!

The Lunar cycle also seems to be directly linked to mating and fertility patterns in humans, as well as governing the movements of fluids. This was confirmed by the work of Dr. Jonas, a Chez physician who was researching female fertility, and who subsequently discovered that human females could spontaneously release an ovum once a month, regardless of their normal menstrual cycle, at the exact moment that the Sun and Moon formed the same angle as at the woman's birth.

For instance, if the woman in question was born at the New Moon, each month when the Moon was new, ie: at the same angle to the Sun as the moment of her birth - she could spontaneously release an ovum, regardless of her normal fertility and menstrual cycles. He then developed a system of birth control based on the combination of the normal menstrual cycle and this extra cycle, which is 97% effective. (You can find the position of the sun/moon angle for your own birth date by referring to an Ephemeris for the year of your birth or consulting a competent Astrologer.)

Dr. Jonas also discovered that the Moons position in a woman's birth chart determined, with 97% accuracy, the sex of her children, and that as the Moon transited positive and negative sectors of the Zodiac it would favour either the 'x' or 'y' chromosome magnetically, in alternating periods of approximately 2.5 days. (A perfect way of maintaining an approximate 50-50 ratio of the sexes.)

In Astrology, the Moon is said to rule the sign of Cancer that rules the home, motherhood, and birth!

Further research into solar activity has confirmed

that great civilisations rise and fall in accordance with sun spot activity, and when the Sun's magnetic polarity, which reverses itself at regular intervals over immense periods of time, occur, the electromagnetic discharge is so great that if affects human DNA, causing major evolutionary changes as well as affecting major geophysical upheavals!

Dramatic changes also occur during eclipses, and in his book 'The Book of the Eclipse' author David Ovason correlates great historical events that have been triggered by, and coincide with solar eclipses. He writes - 'It does not matter at all in astrological terms whether the eclipse is seen or not seen. It still influences the entire spiritual atmosphere of the Earth and works upon all sub-lunar creatures equally, according to the nature of their birth.' He continues 'Scientific research has shown that liquids and substances immersed in liquids change in some inexplicable way during an eclipse, just as the moon helps control the tides. The implications of this are extraordinary, bearing in mind that our physical bodies are vertical columns consisting of about 80% water'.

The Japanese researcher Dr. Masaru Emoto goes even further, and in his remarkable book 'The Secret Messages in Water' suggests that water is not just physically responsive to planetary magnetism as Ovason would suggest, but also more remarkably, to the energy of our thoughts. His stunning photographs of water crystals exposed to different thought patterns and environments are truly remarkable. Like auras, they show an endless variety of shapes and colours - positive thoughts, vibrations or environments producing beautiful intricate patterns like snowflakes, and negative energies producing distorted and unpleasant images,

As well as emitting various electromagnetic forces that affect us here on Earth, however, the Planets also emit sound. This was first confirmed when the U.S. Space Probe, Voyager 2 approached Saturn. The probe picked up a distinct noise from the magnetosphere of Saturn, which was then beamed back to Earth. When these noises were speeded up and played through a musical synthesiser, the waves were found to consist of a distinct melody! Later expeditions confirmed that sound was also occurring within the Sun itself which 'sounds like a chorus of instruments playing tones'. Does each of the planets emit musical notes?

Sound recordings made in space by NASA are said to sound like an incredible symphony, and if the Planets are emitting different notes and tones - and we have already seen that music can affect the shape of matter - are we responding at a subatomic level to the loud but silent tones being emitted by the Planets?

Pythagoras rightly observed that the ratio between the notes on the musical scale were identical to the ratios of the distance of one planet to another.To Pythagoras, music was one of the divine sciences and its harmonies were, without question, defined by mathematical proportions. He believed that mathematics demonstrated the way in which the 'creative force' established and maintained universal laws. Number precedes harmony, since it is an absolute law that governs all harmonic proportions. Pythagoras further divided the creation into a vast number of planes or spheres to each of which he assigned a tone, a harmonic interval, a number, a name, a colour and a form.

Having established music as an exact science, Pythagoras applied its law of harmonic intervals to all the phenomena of Nature, even ultimately demonstrating the harmonic relation of the planets, constellations and each of the elements to each other. Pythaorgas, like the Chaldeans, believed that the heavenly bodies joined in a cosmic chant as they moved across the sky. Pythagoras further believed the universe to be an immense monochord with its single string connected at its upper end to absolute spirit, and the other end to absolute matter. In other words, this chord stretched between Heaven and Earth.

The names given to the various notes of the diatonic

scale were also derived from an estimation of the velocity and magnitude of the planetary bodies. Each of these gigantic spheres as they rushed through space was believed to make a specific tone caused by its continuous displacement of the aethereal diffusion. Pythagoreans also believed that the planets emit certain sounds according to their different properties.

In the Middle Ages astronomy and music were grouped together in the *quadrivium* which also included mathematics and geometry. The German astronomer Johannes Kepler drew on this ancient Greek concept of 'the music of the spheres', also known as *'musica universalis'* to map out the planetary systems.

As we saw in the experiments with the Tonoscope, if sound has the ability to organise matter into distinct patterns, and if the Planets are continually emitting sound, then perhaps the resonances, harmonies and vibrations they constantly emit had a creative effect on the primordial Oceans of the Earth, enabling organic structures to arise from the rich biochemical soup of the oceans, by the various musical symphonies which the planets were creating!

In the same way that the musical scale, composed as it is of a limited number of notes, can produce an infinite variety of musical compositions, the planetary 'tones' may well define and 'orchestrate' life systems on Earth, by their direct effect on the organisation of matter into distinct patterns and shapes. Each planet would have a job to perform, as each note has a job to perform in a chord, a song or a symphony! Each of the inner seven planets are esoterically associated with one note of the octave, the outer planets forming the higher octave. The sun, the centre of the solar system, is associated with middle C.

Michel Gauquelin, the French researcher, confirmed the influence of planetary bodies over individuals when he examined the birth charts and professional occupations of thousands of volunteers, which proved that the dominant planet in the individual chart coincided with professional occupation. For example, sportsmen and soldiers had a prominent Mars in their birth charts, scientists a prominent Saturn, and writers prominent Mercury. (As a writer, I conform to the latter.)

Scientists now confirm that there is growing evidence of birthdays influencing health and careers. Research conducted by Professor David Phillips at the University of Southampton has proved that babies born at different times of the year are more likely to develop certain traits. Of course, they are reluctant to accept astrological explanations, saying 'there must be some scientific explanation.'

Michel Gauquelin believes these varying but predictable influences are the results of terrestrial magnetic disturbances that worked because the solar system was a 'great organism composed of many forces which interact on many levels. The most plausible theory being that the sun is the motor and the solar field is the medium. The moon and the closest and largest planets to the sun cause agitation in this field, and the stronger the agitation, the greater the effect on the child at birth'.

Can this scientifically precise correlation between heavenly and earthly events linking the macro and microcosm - 'as above, so below' - be explained?

In The Hermetica, attributed to the Egyptian sage Thoth - otherwise known as Trismegistus meaning 'thrice great', such was his wisdom - Thoth writes that the creator - Atum - arranged the constellations of the Zodiac in harmony with the movements of Nature saying, 'I will build the Zodiac - a secret mechanism in the stars, linked to unerring and inevitable fate. The lives of men, from birth to final destruction shall be controlled by the hidden workings of this mechanism'.

Is sound the key to this mechanism? Is 'The Music of the Spheres' a literal reality representing precise universal laws that control the creation and evolution of life on Earth at an atomic level, as both astrology and vibrating superstring theory might suggest?

The planets are undoubtedly the largest collection of atoms and vibrating superstrings, and therefore the loudest of instruments in our vicinity. They create the music of the spheres with their notes and tones. They are instruments and the orchestra. And we are all dancing to their tune!

No tune in living history, however, ever wrote itself. A tune must have a writer, a creator, or a composer.

Who wrote the 'Music of the Spheres?'

Chapter Two.
THE HUMAN MIND.

The Invisible Universe.

'And God made man in his own image'.
Genesis 1:26

In the Old Testament of The Bible it tells us that 'God made man in his own image'. What does this mean? We have already determined that this 'God' - the invisible and divine source of all creation - must have existed in some higher mystical form before material creation itself. We have also determined that this God is essentially non-material by nature - an indescribable form of some higher force most closely resembling pure energy, consciousness or light - and which contains both male and female energies (Yin and Yang) simultaneously.

If then, we are not made in God's physical image, to what can this statement refer? Could it mean that we are a microcosmic form of this same energy, consciousness or light from which God made all things? Is this what Christ meant when he said 'I am the Light'?

In the words of the Sufi mystic Kabir, 'All know the drop merges into the Ocean, few know the Ocean merges into the drop.' This statement implies a divine omnipresence, the macrocosm within the microcosm. God within all form.

In the ancient Egyptian texts of The Hermetica it is also written that God is Oneness and that everything is part of one Supreme Being - an invisible universal consciousness, or mind - which has created all things from an original impulse, and yet which also resides within all of creation simultaneously! What relationship does the human mind have to the universal mind? Are our minds really part of an interconnected 'divine consciousness', or are they individual and separate, merely the result of the activity of the physical brain?

It is, I believe, entirely an aberration, ignorance and arrogance of western materialistic scientific thinking to assume that 'consciousness' or 'mind' is a quality possessed by human beings alone, or that this 'awareness' is entirely a function or by-product of the physical brain.

The brain, like all physical body parts and material forms, is absolutely finite and depends on this self same omnipresent universal energy for its very existence. By this simple fact alone, the brain itself cannot be considered capable of producing thought or consciousness per se - because it is ultimately a finite material form and like all finite material forms cannot have either life or consciousness without energy. Rather, it may be that the brain acts as a medium by which 'consciousness' or thought expresses itself on the material plane.

In this view, the brain is considered a physical organ that can only be activated via a continuous series of electrical impulses. Once you deprive the brain of these electrical impulses as in death, the brain will automatically stop functioning. It may remain perfectly intact as a physical mass, but it is the electrical energy that both creates and stimulates it. Switch off the electrics - and the brain is nothing more than a jelly like mass of grey matter waiting for an electrical fix - like the monster Frankenstein waiting for the gift of life from a lightning bolt.

Therefore, even though the brain may survive physical death intact, in exactly the same state it was in just before death, it is no longer of any use - it is no longer activated. The electrical force that once generated activity in the brain itself has withdrawn with the original organising energy field that created it in the first place.

Although Western medical opinion is generally of the

belief that it is the brain which secretes thoughts 'like a liver secretes bile' current evidence is beginning to prove otherwise.

As we shall see later during our investigations into the ever-increasing reports of the out of body and near death experience, the 'consciousness' of the individual appears fully able to separate from the physical body itself during life and describe distant locations - and even more remarkably - also appears to survive and be able to communicate after physical death itself. If consciousness were truly restricted to the physical brain and brain function, it would be impossible for anyone to do this.

According to the concept of divine omnipresence, which suggests that consciousness is inherent in all forms, the physical body is said to contain this inherent energy field that is simultaneously the source and mind of the form itself, whether it is a rock, an animal, a plant or a human. All life forms, in this philosophy are equal and equally endowed with divine universal 'consciousness'. This is the meaning of omnipresence. It means that God is not separate from creation, but intrinsic within it, and that God is everywhere and everything simultaneously.

It is only in our human arrogance that we assume that we are unique in the possession of 'consciousness'. But how does consciousness function in a form without a brain, and the five senses as we know them? A plant for instance - how can a plant have consciousness? It has no brain! It has no eyes, ears or nose! And yet scientific experiments confirm that plants can have awareness of their environments and can suffer emotional extremes of happiness and pain in the same ways that we can. How a carrot must cry then when it is picked, no less than the cry of a fully universally conscious lamb heading for the slaughter.

Man, carrot and lamb all contain this same universal substance - this same universal consciousness. Just because carrots and lambs do not share our interest or skills in computer technology, literature, fashion or space travel is no reason to assume they are not conscious or do not possess 'awareness' 'feeling' or 'consciousness'.

In 'The Secret Life of Plants' authors Peter Tompkins and Christopher Bird describe scientifically controlled experiments where plants were either actually hurt, or the thought of the intent to hurt focused on them. In all cases, the plants - which were attached to extremely sensitive electrodes - showed an extreme emotional reaction whether they were actually hurt, or merely had the thought - the intent - projected at them!

In a series of further experiments which the authors describe in ample detail, a further selection of plants were either physically or mentally attacked by an individual unknown to the researchers. The researchers were later able to identify the attacker from a large number of possible suspects - entirely by the reaction of the subject plant when it encountered the aggressor! Conversely, as any gardener or green -fingered soul will tell you, plants respond extremely well to love and affection!

This research data implies that plants not only react to various stimuli in a conscious and emotional way, but are also able to read our thoughts as in the case of the plants who telepathically reacted to aggressive mental projection!

How can they do this? Now, plants may not have a brain, but they are absolutely capable of consciousness because they contain this omnipresent energy like everything else in the universe and it is this self same energy that possesses consciousness, not the form itself. (The consciousness, spirits or souls of plants have long been known in Celtic Mythology as the Devas of the plant kingdoms - fairies, pixies, and elves who are said to be endowed with special powers - The Oak tree, for instance, is one of the 12 sacred trees venerated by the Druids and Pagan Britons. Socrates called it 'The Tree of Wisdom'. Why were these trees considered sacred? How could they have influence or consciousness?)

Now, because everything in the universe is composed entirely of atoms, and because these atoms contain both a positive and negative electrical charge, the resultant effect is that every single physical form in the universe is possessed of an electromagnetic force field that surrounds and interpenetrates the object itself. What does this force field look like?

In 1950, two Russian scientists, Simyon and Valerina Kirlian discovered that they could scientifically photograph a strange luminescence which appeared to surround and interpenetrate all living things, rocks, stones, plants, animals and humans alike.By laying film or plate in contact with the object to be photographed and passing through the object an electric current from a high frequency spark generator which put out 75,000 to 200,000 electrical pulses per second, the whole body of the physical object in question appeared to be suffused with an inner light which radiated in varying colours and intensities around the object - fluctuating with internal and external changes.

The Kirlians started by photographing plants, which seemed to emit showers of different coloured lights from channels in their leaves and progressed to humans who also appeared to have these fine sheaths of light emanating from them. They also subsequently discovered that the energy field - the field of light - would display positive or negative changes on the energy field itself before these changes became apparent on the physical form.

Two physically identical leaves for example - when subjected to Kirlian photography - showed that although they appeared thus to normal vision, one of them actually had a very strong luminous auric field, whereas the energy field of the other was extremely weak. The latter soon became sick and died, showing the same irregularities on the physical form as it had already displayed on the auric field. The former continued to live and retained its healthy glow.

The same phenomenon has been proved with the study of human energy fields. The field of preventative medicine could surely benefit enormously from such technology!

In all life forms, it is the quality and quantity of energy within the form itself that determines its health and strength. It is the energy within the form that is affected by internal emotional and physical factors, as well as external electromagnetic forces that are being received from other life forms.

The human energy field is also affected by the magnetic fields of the planets that are being continually absorbed by the Earth's magnetic field. These fluctuations in magnetic energy are then detected on the auric field as a disturbance in wave energy.

Plants, animals, humans, even rocks, all are continually giving out positive or negative energy fields, depending on its own state of health and being. They all have their own electromagnetic fields that are all interacting with each other at the sub-atomic level.

In 1953 Dr. Mikhail Kuzmich Gaikin a surgeon from Leningrad confirmed that the auric photographs the Kirlians had produced showed identical points where Vital Energy, of Life Force could be tapped. These coincided with the 700 acupuncture points he had learnt about from Chinese doctors during World War II.

The Earth itself is also surrounded by an enormous vital energy or electromagnetic field of its own, and this magnetic shell is sensitive to all the variations in electromagnetic energy emitted by the Sun, the planets and the Moon, which is the closest orbiting body to the Earth, and the Earth's magnetosphere is affected accordingly.

In a threefold action, the planetary fields bombard Earth with electromagnetic energy, the force field of Jupiter for instance, is 6,000 times greater than the Earths and a hundred times larger, which in turn influences the Earth's magnetic field, which, in its turn, affects the electromagnetic force fields of all life forms on Earth, including ourselves!

The planetary bodies are massive generators of sound and electromagnetic energy, and as they orbit the heavens, forming various angular relationships with each other and us, the varying combinations and frequencies of energy have a direct influence on life on Earth.

The force field of each individual is linked to the universe via the bioplasmic body and the electromagnetic field of the Earth, and the Earth magnetically acts as a receiver for all incoming electromagnetic wave energy. When there is a change in the universal energy matrix, a resonance is produced which then distorts the bioplasmic body of the Earth, which in turn affects the bioplasmic bodies of all life forms on Earth.

In 1959 Dr Leonard Ravitz of William and Mary University in the United States, demonstrated that these human energy fields fluctuated with mental and emotional activity, (as the Kirlians had discovered), and that the resultant wave energy emitted varied in brilliance, intensity and colour according to the psychological state of the person concerned.

Inside the electromagnetic force field, or bioplasmic body, processes appear to have their own energy pathways or ley lines, and the bioplasmic body itself appears to be a unified organism which acts as a whole, is polarised, gives off its own electromagnetic fields, and is the basis, or blueprint for the subsequent biological, or material form.

Later scientific research confirmed that this electrical force could be measured with a scientific device known as a Superconducting Quantum Interference Device (SQUID) that was able to measure the electrical activity of brain function, by the variations in this electromagnetic force field, in even more detail than the EEG.

Twenty years later, Dr Robert Becker of the Upstate Medical School in New York confirmed the existence of these complex electrical fields within the human body and found that they changed in strength and intensity depending on psychological factors. He also discovered electron-sized particles moving through this electrical field. (Have you ever walked into a room and instantly felt a vibration, an 'atmosphere'? You may be sensing the state of the magnetic energy of the environment or person in question!)

Dr. Rupert Sheldrake, a well-known author and researcher who has worked extensively on the morphic (electromagnetic) fields of animals believes that 'all life-systems are regulated by invisible organising fields of energy'. 'These fields', he believes, 'are causative of the physical form and act as a blueprint for the individual form itself'.

Dr Sheldrake's own research has reinforced his belief that these bioplasmic, or morphic fields, could have an effect on other morphic fields regardless of how near or far they were. His extensive research into morphic fields and animal behaviour also shows that animals are extremely sensitive to these fields and are able to experience telepathic sensitivity to their owners, even at great distances.

According to current scientific thinking, the world of apparently solid form is, in reality, suffused with an invisible world of radiating energy fields, thought fields and bioplasmic forms. Furthermore, the mind and body are now being redefined in terms of energy impulses and rhythms, all of which are in constant motion, and each of which can have a direct effect on the magnetic fields of other bodies, however great or small.

Dr Victor Inyushin of Kazakh University in Russia researched human energy fields over many years, and his results suggested that the electromagnetic force field was composed entirely of ions, free protons and free electrons, which was a state absolutely distinct from the four known states of matter, solid, liquid, gas and plasma. The bioplasmic (electromagnetic) force field he subsequently defined as the 5th state of matter.

He also discovered that despite the apparent stability of the bioplasmic field, a large amount of energy was radiated out into space through breakaway particles that could be measured as they emanated from the bioplasmic body.

According to his extensive research, which began in 1950, the bioplasmic field appeared to be a balance of positive and negative particles within the bioplasmic field and was relatively stable. If, however, there were a shift in the positive and negative balance of the bioplasmic body, ill health would be the result.

What causes the bioplasmic body in humans to fluctuate - what is the agitating factor? The agitating factor is thought. Where does thought come from? It comes from our minds.

What is the mind? What, ultimately, is consciousness?

The Aura - Soul Within

'Let thine sight be single and thine whole body shall be full of Light.'

Mathew 6:22

One of the most common recurring themes in all the world's religions is that humans have a divine immortal soul, which survives physical death. We know, of course, that our bodies will eventually die and disintegrate because they are made from finite matter. So, what exactly is the Soul? How can it survive physical death when we know our bodies will simply turn back to dust?

Quite simply, the soul is the orb of conscious energy 'The Light' within us. It survives physical death because it is not physical. It is pure energy. Pure consciousness. Pure Light. It is non-material, infinite and indestructible.

The 'light energy', or consciousness, which is simultaneously the cause and mind of all things that exist, and which will survive intact after it has withdrawn from the physical body at death, is what we know as the soul. The individual soul however, although part of the greater Oneness, also retains a perfect record of its own eternal existence, (as we shall see later in the chapter on reincarnation).

Ancient mystics have long described an ethereal energy body, which duplicates the human body, and which is perhaps best defined as an electro-magnetic area where subatomic vortices of the cosmos are transformed into the individual.

The ancients also believed that the 'aura' was an exact double of the physical body but composed of a much finer substance. The aura is also said to contain all the information about the individual from past, present and future.

In ancient Egypt, this spirit double was simply known as 'Ka'. The Azande tribe in Africa believed that everyone has an etheric double, which they called the 'Mbisimo' - a mirror of the physical body, but composed of much finer matter. The Bacairis of South America talk of a 'Shadow' that leaves the body during sleep and finally at death, and in Burma, the 'spirit double' is simply called 'The Butterfly'.

That plants, animals and humans have fine sheaths of subatomic or protoplasmic energy interpenetrating the solid physical bodies of molecules and atoms is a concept that dates back for thousands of years, long before the Kirlians invented their ingenious camera.

Although normally invisible to the naked eye, the aura usually appears as a large circle of light surrounding the physical body and can extend out from the physical body from a few inches, in the case of a weak auric field, to an average of about three feet to five feet in a healthy person. The especially trained or highly spiritually developed such as the Yogis of India, can have an auric field in excess of ten to thirty feet or more. The aura of the Buddha was said to stretch for several miles!

The aura, or halo of light that surrounds and

interpenetrates the physical body is also well depicted in Christian art, in the ancient iconography, and in the stained glass windows of churches and cathedrals, the aura of light - or 'halo' of light - appearing around the heads of saints, and even Christ himself.

However, the halo or aura of light in reality actually surrounds, not just the head, but the whole body. The symbolism of the aura depicted around the head in iconography was simply meant to indicate those who had become spiritually illuminated or highly spiritually evolved.

In the Gospel of St John, however, this 'Light' is referred to in even greater depth: 'There was a man sent from God whose name was John. This man came to bear witness of the Light that all through him might believe. He was not that Light, but was sent to bear witness of that Light. That was the true Light which gives light to every man who comes into the world.'(John I: 7-9). In Thessalonians 5:5 we are told 'You are all sons of Light'.

The theme of light occurs in many religious writings and in part 3, verse 2 of the Mundaka Upanishad: (essential scriptures in Hindu Vedic philosophy) we read:

'As long as we think we are the ego,
We feel attached and fall into sorrow.
But realise that you are the Self, the Lord Of Life,
And you will be freed from sorrow.
When you realise that you are the Self,
Supreme source of light,
Supreme source of love,
You transcend the duality of life
And enter into the unitive state.'

Twentieth century parapsychologists have also come to view man as linked to the universe via his auric body, and now believe that we all react to changes in the planets as well as to the moods, thoughts, and emotions of others. We are also affected by sound, light, colour, magnetic fields, weather and noise. When there is a change in the universe, a resonance is produced in the vital energy of the electromagnetic body, which in turn affects the physical body.

Auric Healing.

'Although mystics have not spoken of energy fields or bioplasmic forms, their traditions over 5,000 years in all the parts of the globe are consistent with the observations scientists have recently begun to make'.
'Hands of Light'. Barbara Ann Brennan

The ancient Chinese in the 3rd Millennium B.C. believed that everything was composed of ' Ch'i,' or vital energy, and that Ch'i itself was composed of two opposite but complementary forces, Yin and Yang. When Yin and Yang were in balance, the subject experienced good health, but when out of balance, ill health resulted.

They subsequently developed acupuncture, a system of healing which addressed this inner energy, and by manipulating the life force, as it ran along the invisible energy lines within the body, they could help restore the balance of positive and negative energies.

In its original sense, the word healing means to complete, 'to make whole'. To achieve a state of health or wholeness, a balance must be found between these two positive and negative energies. When we are out of balance, ill health is the ultimate result.

The aura itself can best be described as pure energy, or light that is primarily agitated by thought vibrations. Eventually, however, the thought vibrations crystalise and densify on the physical body.

A trained energy healer or spiritual healer will be able to detect these imbalances on the auric field, before they manifest onto the physical body. Those gifted with clairvoyant sight will be able to detect them visually by using the rods and cones in the corner of the eye - peripheral vision - which are more sensitive to subtle energies and wavelengths. When in the process of energy (spiritual) healing, the electrical field of the healer increases drastically in voltage as the healer passes energy on to the patient!

Verena Davidson, a gifted spiritual energy healer, claims that when in the process of healing 'I tune into a higher source and simply ask to be a clear channel of universal energy. I then imagine that same energy radiating out of my hands and I find I can alter energy fields and vibrations in this way. Working primarily through the seven chakras, or wheels of energy, I am able to detect energy blockages or imbalances. Because each chakra resonates at a certain tone and channels one of the colours of the spectrum, I am able to calm or energise the chakra, and increase, decrease or cleanse the colours there.' The hands, she says, act as a scanner, picking up energy from the patient.

More new-age touchy-feely mumbo-jumbo? Not so! Scientifically verifiable!

Dr. Robert Beck is a nuclear physicist who has travelled the world measuring the brain waves of healers in an attempt to try and explain the healing force. Dr. Beck soon made a startling scientific discovery. He discovered that no matter what their own particular method of healing, all healers exhibit an identical brain wave frequency during healing itself. The frequency was 7.8 - 8.0 hertz! Why should it align to this frequency? This puzzled Dr. Beck until he realised that the Earth's Magnetic field also fluctuated between 7.8 - 8.0 hertz! These are known as Schuman waves.

Dr. Beck also discovered that during the healing process the healer's brain becomes both frequency and phase synchronistic to these waves and the natural frequency of the global magnetic field. Dr. Beck also believes that healers are able to take energy directly from the magnetic field of the Earth itself at the unconscious level. This phenomenon is known as field coupling. According to Dr. Beck's research, after healing the patient's brain frequencies had also become phase synchronistic with the healer and the 7.8 -8.0 hertz frequency of the Earth's magnetic field.

Dr John Zimmerman, founder of the Bio-Electro Magnetics Institute in Reno, Nevada, also researched this field coupling phenomena and found that the healers left and right brain became balanced with each other and demonstrated the same Alpha rhythm predominating during healing as Dr. Beck had found.

In our original form we are pure energy and pure light and beyond sickness and ill health. It is negative thinking that causes imbalances. Ultimately, everything is pure energy, and if we are not in balance and living in harmony with ourselves, and the universe, these negative emanations will eventually materialise onto the physical body and begin to block the energy fields resulting in ill health.

Natural healing methods help direct healing energy to the auric field and help correct these imbalances. The type of imbalance may also be detected by assessing the colour within the energy field itself. Whatever colour or chakra energy is out of balance can be rebalanced, cleared and energised.

We are continually creating colour within our thoughts which filter onto the auric field, and in constant reaction with colour from our environment. Although we may think of colour as an inert substance with no power or meaning, colour can affect our mood and temperament. We also use colour to express our feelings - thus, 'red with anger', 'green with envy', and so on. Can colour affect our mood?

In 1932 Robert Gerard, a US scientist conducted a series of scientific experiments. He subsequently discovered, under strict laboratory conditions, that exposure to red light excited the senses, increased heart and respiration rate and inevitably made people much more stimulated and aggressive. Exposed to pink light however, his experimental subjects became tranquil and incapable of anger! Blue too, was calming and relaxing, decreasing pulse and respiration rate and so on.

These effects, he noted, applied to those with or without physical sight! This, one might assume, is because light behaves as both particles and as waves, and can be perceived with both physical sight, and at the sub-atomic level on the auric field that will be perceived as vibrations, or wave energy, through the relevant Chakra.

We literally perceive colour at the wave level, the auric field acting as a magnetic receptor to wave energy. Each colour has its own wave frequency. We pick up colour at this level in the same way that a radio tunes into radio waves. The aura is continually reacting to colour vibration.

What are the Chakras? According to the Indian Vedic tradition, the Chakras are seven major invisible wheels, or vortexes of energy within the auric field itself. These are situated parallel to the spinal column, and are the invisible receptors for cosmic or universal energy. Each is tuned to a different planetary magnetism, note and colour and twinned with a major glandular system.

The Chakras essentially act as receivers for universal wave energy, each operating at its own frequency of sound and colour. The Chakras then transform this powerful universal energy into an energy vibration the physical body can withstand, via the seven major glandular systems. The glandular systems are then responsible for distributing this energy throughout the body in order to maintain and nourish vital life systems at the physical level.

There are seven major Chakras and seven major glandular systems. Third Eye Centre, or Brow Chakra is paired with the Pineal Gland, the Throat Chakra with the Thyroid gland, The Solar Plexus with the Adrenals, the Bass Chakra with the Gonads, and so on. A disturbance or blockage at one of these paired chakric-glandular centres will cause an imbalance in the energy field and if it is not prevented, will eventually manifest as physical disease.

Colour has a profound effect on our biology at the deepest levels of our being and the colours that appear on the aura are associated with the energy generated and received through the chakras. Each chakra is further associated with a colour, a note, and a planetary ruler, and the colour manifesting on the aura is indicative of mood, spiritual and mental state, and physical health. The stronger the aura, the stronger the light, the stronger the energy. The weaker the light - or life force - the weaker the energy. Blockages on the physical body will be preceded by blockages on the aura itself.

The aura, being as it is composed of some form of sub-atomic vibrating substance, is also sensitive to sound as a healing medium and 'overtoning' is now a popular method of inducing healing onto the auric field by the use of sound vibrations. Each of the seven notes have an affinity with one of the seven colours of the spectrum, and individual notes can energise or calm the various chakric centres.

As well as being indicative of physical health, the colours on the aura are also indicative of personality type. One way of quickly and easily seeing this on the aura is through a technique called Aura Imaging, a recent development of Kirlian photography, to produce printouts of the human aura, together with a personality interpretation. The data is gathered using a special handset that measures the energy in and around the body through the hand.

Aura imaging researchers have found a strong correlation between different colours in the aura and personality attributes and states. For example red is the most physical and action orientated, green is about communication balance and nature, violet is intuitive artistic and visionary, yellow is analytical and intellectual, and so on. Different colour types also have different ways of staying energised or balanced.

At the invisible level, there are many forms of healing

and many psychological tools we can use to improve our health and wellbeing. From Kirlian photography, to aura imaging, spiritual healing, aromatherapy, homoeopathy, acupuncture, herbal medicine, sound therapy - all these have a direct effect on the vibration of the auric field of the individual and assist the re-balancing of the self and encourage the healing process.

Although modern medicine and the medical establishment remain hostile towards such 'alternative' methods of healing, it should be remembered that our modern system of medicine - so-called 'orthodox' medicine - is a system based on the use of chemical drugs and corrective surgery, and is actually less than 200 years old. Acupuncture, herbal medicine, energy and colour healing however, are mentioned in ancient texts from all around the world, North and South America, Africa, Australia, Ancient Egypt, India, China, Peru, Mexico - natural healing may even go back into pre-recorded history by as much as 60,000 years!

Whilst over 20,000 people die in the UK annually from the side effects of chemical drugs - I have no record or knowledge in my research material whatsoever, of anyone suffering such fatalities with natural or alternative medicine.

Furthermore, recent clinical trials conducted in Germany found that the herbal remedy St. John's Wort, a popular herbal folk remedy for depression, worked as effectively as any chemical drug for the condition, many of which have extreme side effects. Researchers from Nottingham University hospital also found that gum from a tree on an Aegean island can cure stomach ulcers and stop their recurrence! Many other ancient herbal remedies are now coming out very favourably in strict clinical trials.

The obvious applications of natural medicine are profound, and today, more and more people are turning to 'alternative' therapies. Of course, there is room in holistic medicine for both ancient and modern medical techniques and I believe that they should work together for the good of humanity, and to take a joint responsibility for its health and welfare.

Our holistic selves, however, are in reality the sum total of the energy apparent on the pre-physical self, and this energy is in a constant state of vibration and fluctuation. The primary agitating factor involved in this process of dynamic change is thought. As we have already seen, thought is the initiator of all things, attitudes, beliefs, habits, and actions.

Whatever we think has a direct effect on our vibrations, and this energy in turn, has a direct effect on the physical body. By realising that negative thoughts create negative actions - which in turn create negative energy which often resides on the physical body, stored in the cells and tissues the self - the thinker is enabled to create, literally positive 'new vibrations' for him or her self, which, in turn, will filter onto the physical body as an improved sense of health and well being.

Conscious and Unconscious.

'The general theory of relativity shows us that our minds follow different rules than the real world does. A rational mind, based on the impressions that it receives from its limited perspective, forms structures which thereafter determine what it further will and will not accept freely. From that point on, regardless of how the real world actually operates, the rational mind, following its self imposed rules, tries to superimpose on the real world its own version of what must be.'

Gary Zukav. 'The Dancing Wu Li Masters'.

Although we now know that these auric or electromagnetic fields are a reality, and they have now been scientifically proven to exist beyond a shadow of doubt, you and I generally have no conscious awareness whatsoever of their existence.

These phenomena occur constantly as an ongoing reality and yet we are normally almost entirely ignorant of them. In the same way that we have no conscious knowledge of this inner world, we are also equally unconscious of other complex and important functions that are a prerequisite to our existence.

An innumerable amount of functions are taking place within us at this very moment of which we have no conscious knowledge. Somewhere, within the self there is a knowledge and consciousness which exceeds our own individual conscious awareness and which is entirely and totally responsible for the day-to-day running of our very being at the atomic and even subatomic level.

Now, whereas the conscious mind deals primarily with what we are 'conscious' of in our daily lives, as interpreted by the five external senses, sight, sound, touch, and smell, the unconscious deals with the complexities of our inner worlds whilst we remain blissfully unaware.

Carl Jung describes the self as the wholeness that transcends consciousness and it was Jung who first brought the concept of conscious and unconscious to modern psychology. Jung also pointed out that what could not be expressed or understood by the conscious mind, could be comprehended by the unconscious, and that the unconscious communicated to the conscious mind in the language of symbols.

The use of symbolism is apparent in the dream state, which is a direct result of unconscious activity, the cause of which is still little understood by mainstream science and medicine. The symbolism which filters through to the conscious mind however, can also represent organic, cellular and atomic forms, for which the conscious mind has no reference, and which can appear spontaneously in works of art, especially those of a spiritual nature.

The Shri Yantra, for instance, which Hans Jenny discovered with his Tonoscope, and which is a literal visual representation of the sacred word Om and symbolises the moment of creation, was first painted over 3,000 years ago by Buddhist monks long before the Tonoscope was even invented! How did the monks access the magical and meaningful Shri Yantra? Jung claimed these images existed at a level within all humanity, known as the 'collective unconscious'.

Now, we know the conscious mind connects us to our everyday reality and our everyday world but we have also seen that a universe exists within, a body of Light, exists within us all, which contains energies and intelligences for which we have no reference.

Although the conscious and the unconscious are two separate components of the mind, they are by no means equal. Scientific tests also reveal that the conscious mind makes up only one eleventh of the mind itself - whilst the unconscious accounts for the remaining ten-elevenths!

We might compare the conscious mind to the tip of the iceberg, and the unconscious mind to the submerged portion that makes up the major part of it. This inner energy, furthermore, being the creative source of the self, has knowledge of all things and it is via the unconscious mind that the inner self can be contacted.

Left and Right Brain

'When you make the two as One, and the inside like the outside, you will enter the Kingdom of Heaven.'
Jesus Christ. Gnostic Gospel of Thomas

The human brain is actually not one, but two brains, composed of two distinct halves simply known as the 'left-brain' and the 'right-brain'. The left and right brains are connected by a single channel, which is composed of billions of nerve fibres and is known as the Corpus Collosum. The left-brain governs the right side of the body, and deals primarily with intellectual, practical and analytical functions. It is considered Conscious. The right-brain governs the left part of the body and deals primarily with imaginative and intuitive functions. It is considered Unconscious. (If you are left handed, the reverse applies.)

By the evolutionary design process of dividing the human brain in two halves, we observe the replication of an important natural phenomenon that appears consistently in the material dimension, and that is, the pairing of opposite but complementary forces, symbolised in the Chinese symbol Tai Chi, or Yin and Yang.

Male and female, night and day, hot and cold, everything in the material dimension can be considered as consisting of two complementary but opposing forces, and the human brain has been designed and developed to perceive the forces of the universe in two distinct ways. The Left Brain, (Yang) which can be considered solar, active and male deals primarily with the conscious mind, and its associated logical, practical and analytical function. The right brain, which can be considered Lunar, receptive and feminine, (Yin) deals primarily with the unconscious mind, and its associated imaginative, intuitive, and rhythmic functions.

In the late 1990's, researchers Bennett and Shaywitz, using electromagnetic resonance cameras that highlight brain activity, discovered startling differences in brain function in both men and women. In laboratory conditions, experiments were conducted where an equal number of male and female volunteers were given a series of questions and tasks to deal with. Without exception, whilst the men responded virtually entirely with their left-brains to all the experimental stimuli, women responded with both halves of the brain simultaneously!

Men's inability to deal with emotional and intuitive functions (in general) can be seen as a lack of activity in the right brain area. Equally, women tend to use less of the left-brain than men, and are sometimes incapable of fully accessing or utilising the potential of the left-brain functions.

At the Cognitive Psychopharmacology Department at the Institute of Psychiatry in London, Dr. Tonmoy Sharma has discovered (with the use of MRI, or magnetic resonance imaging) that women are less able to perform logical functions like map reading. Whereas the male test cases showed a large increase of activity in both areas of the brain, female responses were weak in both left and right brain. In verbal skills such as expressing feeling, however, the female brain lit up drastically on both sides, and the male test cases showed virtually no activity in comparison. We need to utilise and express both sides of our brain for completion and wholeness but unfortunately, in society today, we have seen a collective suppression and underestimation of right brain activity often expressed directly as a suppression of female qualities, and sometimes of women themselves who in various religions and cultures are still considered 'inferior' to men.

However, the Bennett and Shaywitz research has now confirmed that women function at a more complete, intuitive and sensitive level than their male counterparts - but the emphasis on society today is not on the mystical potential of the 'feminine' right brain, but on intellectual and practical achievement. It is still, in essence, a male dominated left-brain society.

The activity of the brain in both sexes however, can primarily be quantified as consisting largely of electrical impulses which are generated on the electromagnetic field before being processed by the brain, which acts as a medium between energy and matter. There are four main modes, or levels of consciousness, each of which operates on a different frequency, or wave band.

The wave associated with normal waking consciousness is the Beta rhythm that operates on a frequency between approximately 13 and 30 HZ (cycles per second). The second wave is known as the Alpha rhythm that operates on a frequency between 8 and 13 HZ and is associated with meditation and relaxation. The third wave is known as The Theta rhythm that operates on a frequency between 4 and 7 HZ and is associated with the dream state. The fourth is the Delta rhythm, which is the slowest and which operates up to 4 HZ. The Delta rhythm is associated with deep sleep.

In the US, Dr Daniel Kirsch has been a leading pioneer in the field of electro-medicine since 1972 and has subsequently devised the Alpha Sim technology which by harnessing neurons in the brain can attune the four brain waves to their correct frequencies, thus reducing medical symptoms such as pain, insomnia and stress. In essence, the human mind is sensitive to electrical stimuli from both within the organism itself, from other objects and beings, and from the external universe. It operates entirely on energy and is affected by all other energy sources.

Is the human mind really a part of one all encompassing universal mind which is composed of some higher energy most closely resembling pure consciousness, energy or light? If so, our minds could contain unlimited power.

Free from the constraints of the physical body, the mind, the soul within us all, which is eternal light and all-powerful energy can merge with the universal mind whose power of creativity is limitless. Connected and interconnected to all things, our individual 'minds' retain and possess a collective universal knowledge that goes beyond the limits of our everyday reality, the definition of which has been imposed by our current limited understanding.

Can the interconnectedness of all things explain much of the mystical and supernatural that we normally reject as 'irrational', 'illogical' or unscientific?

With the exception of Islam (where it is believed God creates the Universe but is somehow separate from it) the idea of 'Spiritual Oneness' or 'Omnipresence' exists as a fundamental belief in Hinduism, Buddhism, Judaism and Christianity. It is only now, however, that science is seeing the parallels in psychics.

At the University of Dortmund in Germany, Theoretical Physicist Heinrich Pas, author of 'The One: How an Ancient Idea Holds the Future of Physics' has come to believe that the ancient idea of 'Monism' is in fact a great scientific breakthrough in searching for a 'theory of everything'.

Basically, Monism postulates that although the universe appears to be made of different bits, it is in fact only one single all encompassing thing, and that everything else we see around us is some kind of illusion. In other words if you apply the concept of quantum entanglement to the entire universe you end up with Heraclitus dogma 'from all things One'.

In the Hasidic view, 'there is nothing but God'. Simply put, if God creates the Universe, before God created the Universe there was nothing but God, so if God creates the Universe it must be God made manifest, because there is nothing else, not before and not after. In the Shema Yisrael, a Jewish Prayer that encapsulates the essence of Judaism, it is written 'Hear (Oh Ysrael) YHWH is our God, YHWH is One.' (Deuteronomy 6:4)

The idea that we are all 'One' is essential in understanding the altered states that appear to defy our current understanding of space and time and indeed consciousness itself. Could these experiences be quantum in nature?

In the following chapters we shall continue to investigate the supernatural and paranormal powers of the human mind, but it is through clinical hypnosis that we firstly encounter and access these deeper levels of consciousness and it is here that we begin to see the mind unfolding.

Chapter Three.
MIND OVER MATTER.

Hypnosis.

'When you go into a trance, you will experience a change in awareness. Just as everybody is different in the way they experience life, so each person's way of experiencing trance is unique.'
    Paul McKenna. 'The Hypnotic World of Paul McKenna'.

Let us firstly imagine that the totality of our consciousness is like a vast body of water eleven miles deep. In this analogy, our conscious awareness would span the surface of the water, and a little under a mile down. The other ten miles deep represents our unconscious minds.

Alternatively, let us imagine that the totality of our 'consciousness' is like an orange. Our conscious minds can be compared to the skin, with its view of the surface and the external universe, and the inner orange represents our minds at the unconscious level, the totality of that which exists invisibly within us all.

In our 'normal' daily 'awake' state of awareness or consciousness, we are only aware of the surface of 'reality' that we perceive with our five senses and our conscious minds. However, we know that there exists within us all a non-physical reality, an auric field, which is both consciousness and energy and which appears to contain all the information regarding the individual soul and its form, and which simultaneously transcends both space and time as we know it.

Of course, we are generally oblivious and ignorant of all this memory and experience in our daily lives, concentrating as we do on our never-ending quest for survival. Nonetheless, this information exists within each of us at a universal and personal level whether we know it or not.

There may be times in each of our lives when this 'ultra-reality' breaks through to the conscious mind in flashes of insight, sudden awareness, a gut feeling, a vision or dream, but there are also many tried and tested methods of deliberating altering the mode of consciousness so that the individual can perceive this inner world, and thereby access both supernatural experience, data and information normally unavailable to the conscious mind.

How do we unlock the door and access the unconscious?

Hypnosis appears to be one important key. Under hypnosis, we can remember details of events we aren't consciously aware of. Under hypnosis we can have operations, control pain and heal wounds. Under hypnosis we can go back in time to remember our childhood days, and even more remarkably to remember past lives, as we shall see later in the chapter on reincarnation.

Although only recently introduced into modern western psychological practice by Austrian physician Franz Mesmer, who believed that a power similar to magnetism existed in the human body, many ancient cultures exhibited knowledge of trance-induction as a healing medium. The ancient Chinese, the Egyptians, the Indians, the Greeks, the Persians, and the Romans all utilised this knowledge for spiritual and healing purposes. The Hindu Veda describes similar procedures which are over 3,000 years old and which resemble modern hypnotic techniques to a large degree.

How is the hypnotic trance state induced?

Firstly, let us dispel a myth. Anyone can be hypnotised. In fact, we all at some time or another during the day pass into minor hypnotic trances. The simplest examples of when this can occur quite naturally are - when we are staring out of a window, driving down a long stretch of motorway, watching your favourite programme on TV, or simply being engrossed in something pleasant you enjoy doing!

Hopefully, as you read this book, you will also be absorbed enough to notice that you haven't noticed the normal sights and sounds that you would if you weren't focused on doing something else! Quite simply, the greater the focus - the greater the attention - the greater and deeper the trance! We all slip in minor hypnotic trances at regular intervals. Why does this happen? Thoughts, as you know for yourself, are normally in constant motion, jumping from one thing to the next - especially in our ultra busy modern world. At this level of consciousness, we are functioning on the Beta brain rhythm operating at a frequency between approximately 13-30 HZ.

By focusing the mind on just one thing at a time, especially if it is in any way calming or relaxing, like listening to music or sitting quietly and listening to the water flowing in a river, or to the ebb and flow of the tide, our minds automatically start to switch frequencies! We start to drift into the relaxed Alpha rhythm operating on a frequency now between 8 -13 HZ! The Alpha rhythm is associated with meditation and relaxation.

Once we quiet the mind and begin to relax, we automatically begin to induce alpha waves. It is apparent from all the medical data on hypnotic trance induction, that in this state, the conscious mind can be by-passed, and in so doing, access deeper levels of consciousness and reality.

Alpha wave brain rhythm is the key to the altered state. By using relatively simple repetitive words, sounds, or suggestions to create deep relaxation in the subject, the hypnotist can induce an altered state of consciousness in the recipient that we call the hypnotic trance. By diverting the focus of mental attention from the left-brain to the right-brain, from conscious to unconscious, the hypnotist or hypno-therapist can assist the subject in accessing otherwise unavailable parts of memory and consciousness, thus overriding the restrictions and limitations of our conscious minds. For instance, there is well-documented film footage of arachnophobics emerging after one hypnotherapy session tenderly holding a large hairy tarantula. Now, that is truly amazing!

This technique of bypassing the normal conscious responses and tapping into a more powerful aspect of the psyche has been used successfully in all areas of medical and psychological health, and offers a wide range of potential in medical practice. It is only unfortunate that modern medicine does not place more value on these techniques that could prevent untold suffering and conserve vital, and often expensive resources.

Burns victims, for example, are remarkably responsive to early hypnosis to prevent skin damage. Studies also show that in a receptive hypnotic subject even the transferred suggestion that a hot object is placed on the skin will actually produce large painful blisters. This is despite the fact that there is no actual burn, no thermal stimulation. At the same time, suggest that the skin is cool to a burns victim, especially within the first two hours after the burn, and it will prevent the usual development of inflammation.

In all areas of medical practice, there are many applications for Hypnosis. In dentistry, nursing, psychiatry, childbirth, anaesthesia and surgery, hypnosis has, without a doubt, proved that we can transcend the limits of the conscious mind and access a far more powerful form of consciousness. There are also enormous benefits in self-hypnosis.

Bernadine Coady, 58, originally from the West Indies, and the matron of a nursing home in Cambridgeshire, England, successfully completed a diploma course in self-hypnosis from the British School of Hypnosis. Some years later, she was due to have an operation and she decided, for medical reasons, to use hypnosis instead of a general anaesthetic. Unfortunately, on the date in question, her hypnotist failed to materialise. Undeterred, Bernadine decided to draw on her own skills and hypnotised herself. Before surgery she spent just three minutes quietly talking herself into an altered state of mind. She told herself she would feel no pain and that if she did, she would like it to be waves washing against a sea wall. Each time the pain happened, she visualised it ebbing away, like the tide. Bernadine had a successful operation anaesthetised only by the power of her own mind!

How does hypnosis work? Once the conscious mind is sufficiently relaxed, the brain starts to produce alpha waves that automatically induce dream or trance-like states whereby the unconscious can be accessed. Once we access the unconscious, or right brain, we connect to an unlimited reservoir of knowledge, memory and experience. If we are able to tap into the unconscious to such a degree, and bypass the confines of the conscious mind, what is it in fact that we are actually tapping into? As we have already seen, we contain information and data at the unconscious level, and are continually absorbing information of which the conscious mind has no apparent awareness.

To access this immense unconscious data bank of information we have to switch-off the intellectual apparatus, the medium for conscious thought and perception - the left-brain. We have to delude it, literally hypnotise it - lull it into a non-active phase! As soon as we have shut this door, as soon as we put the conscious mind to rest, to sleep, the other door - the door into the unconscious - opens. However, that is not to say that under hypnosis the individual is asleep - far from it. It is the conscious mind that 'goes to sleep' or becomes unconscious during this process, enabling the unconscious to be activated, and the hypnotic subject to access hitherto inaccessible data. The patient is fully aware of what is happening, but on an altogether different level and different mental perspective.

Altered States.

*'To see the world in a grain of sand,
A Heaven in a wild flower,
Hold Infinity in the palm of your hand
And Eternity in an hour'.*

William Blake.

What is the measurement of reality? Do we define reality as the world we perceive when we are conscious, and relating to the externalised material universe, (left brain) or would we be better define it as the world of the imagination and the unconscious (right brain)?

Although we know this inner world to exist, can you see or hear the atoms rotating and vibrating within you? No. You cannot. Certainly not with your conscious mind at least! Can you see or hear the planets humming and creating ripples of energy in space that are distorting your own auric field? No. You cannot - certainly not with your conscious mind at least.

To the five senses, to the conscious mind, a brick wall is always a brick wall. Subject the brick wall to further scrutiny however and you will find that it is made up of billions upon billions of atoms which consist of minute rotating bodies, which, given the nature of rotating atomic bodies, will be vibrating, giving off electromagnetic energy, singing at the superstring level, and shining with a given quantity of inner light.

Our five senses perceive only the brick wall. But we know these things are realities! Science has confirmed it! We know all this invisible activity is going on everywhere all the time! It's a fact! Our very own bodies are full of tiny vibrating atoms that are giving off energy, sound and light and colour. Can we hear or see them with our five senses - our conscious minds? No, we cannot.

So which is the reality, the brick wall, or the vibrating

atoms? There is a spiralling world of energy and light within our own beings that is constantly reacting to levels of consciousness, energy and vibrations that are normally beyond the reach of our everyday reality and our 'normal' perception. Who is to say which is more real? Just because the atoms are invisible to our normal sight does not mean they do not exist.

Ultimately, reality should not imply either one or the other, but rather, of both. We are in effect, equally capable of perceiving two entirely different worlds, the world of matter - external world - and the world of invisible radiating energy - internal world, for, as we have already seen, within every living being in the world, animal, vegetable, or mineral, there is a micro-cosmic world which imitates the creative principles of the entire universe, that is, light materialising itself through ten dimensions (vibrations) until it manifests as physical form.

The universe is essentially light materialised into form and, we too, are microcosmic amounts of light individualised into form. The amount of light (energy) required to make the individual form is appropriate to the size of the form itself.

Although we are unable with the conscious mind to penetrate these inner realities with our intellects and our five senses (left brain) we are enabled to experience something of this inner world of light, energy and higher dimensional realities, by entering into altered states of consciousness (right brain).

In many cases the desire for altered states is primarily a spiritual one that is driven by a need, either conscious or unconscious, for unity with the self. It is nothing less than a desire to transcend the world of duality for the unitive state, and throughout human history, peoples of every race, religion and culture have been compelled to experience something of this altered state, to experience something of this inner universal energy, this inner reality, and to access higher states of consciousness or reality. Perhaps the oldest and earliest of these rituals was achieved by the ingestion of psychoactive or hallucinogenic plant substances.

At some time in our long distant past, human beings first discovered, probably entirely by accident, that certain plants - that certain hallucinogenic plant material - had an inherent ability to alter perception and consciousness in the individual to a profound and remarkable degree. Normal consciousness and daily reality would be overtaken by the experience of increased perception and the ability to see beyond the visible to the normally invisible universe.

The use of hallucinogenic plant material soon became an essential part of religious ritual in our primal societies. Some of these tribal rituals survive to this day in the Rainforests of the Amazon and in remote parts of Africa and Asia. Collecting and grinding special plant material which is either eaten, or ingested through the nasal cavities, the native shaman still connect to the spirit worlds in this way and in their dreams and trances they say they can transcend the physical, material world, and enter into a luminous world of spirits and ancestors.

Today, these ancient Amazonian tribal rituals are being resurrected trans-globally. Ayahuasca, a plant based psychedelic made from Banisteriopsis Caapi and Psychotria Viridis plants, is becoming increasingly popular with those who wish to expand their consciousness, and whatever country you are in there will be those offering to recreate these Shamanic rituals.

Ayahuasca itself contains DMT which is a hallucinogen and which induces changes in the brain involving feeling, memories, vision and which allows for amplified introspection and problem solving related to past and current life stresses. Some say it's like having a lifetime of therapy in just a few hours. Many participants report powerful hallucinations, near-death experiences and journeys through alternative realities.

In recent studies conducted at Imperial College London, scientists recorded the participants' brain activity before, during and after the drug took hold. The MRI images showed that the brain's normal organisation breaks down, and connectivity between regions soars, no doubt across both the left and right brains.

Even Prince Harry has now publicly admitted using Ayahuasca and said it helped him 'deal with the traumas and pains of the past, bringing a sense of relaxation, relief, comfort and lightness.'

The positive power of psychoactive plant material is not arbitrary however, or confined to one study, and is now being proven effective in other clinical trials where renewed investigations are taking place on the use of psychedelic substances for treating addiction, depression, anxiety and PTSD. The hallucinogens it appears, can reach parts of our consciousness that are otherwise inaccessible. Most however, remain illegal.

Today, attitudes are changing, and in the USA, further clinical trials are underway using psychoactives to help treat everything from depression to PTSD. At the John Hopkins centre in the USA, they are having very impressive results showing an almost 80% improvement in the patients concerned after controlled therapy known as Psychedelic Assisted Therapy.

Apparently, psychedelics like Peyote and MDMA - as was found with the studies into Ayahuasca - enable the brain to expand its connectivity and enable patients to break sub-optimal patterns to forge new neural pathways. Diagrams of before and after PAT show a remarkable increase in brain connectivity.

Used since humanity first evolved for spiritual cleansing and connectivity to ourselves and nature, our Western societies are now seeing that there are medicinal benefits to these psychoactive plants that go far beyond recreational use.

This desire to transcend the limitations of the perceptual world has remained in our genetic makeup and remains within the psyche, and within our natural cultures to this day. Psychologist David Clark believes that the massive social shift and revival of the use of hallucinogenic drugs in the last thirty or forty years is an unconscious attempt by the collective unconscious to reconnect to these inner spiritual realities, in search of a lost spirituality and psychic wholeness.

In his book 'The Doors of Perception' the writer Aldous Huxley - who spent half his own adult life researching the psychedelic experience wrote 'I continued to look at the flowers, and in their living light I seemed to detect the qualitative equivalent of breathing - but of a breathing without returns to a starting point, with no recurrent ebbs, but only a repeated flow from beauty to heightened beauty, from deeper to ever deeper meaning. Words like Grace and Transfiguration came to my mind. The beatific Vision, Sat Chit Ananda, Being - Awareness - Bliss -for the first time I understood, not on the verbal level, but precisely and completely what those prodigious syllables referred to. I was in a world where everything was suffused with an inner light.'

As a writer of some repute, Huxley was able to describe something of his experiences on Mescalin where he spoke of a world of radiating light, of a beauty beyond words, a world transcendent from the material dimension. Somehow it would appear that hallucinogenic plant material can bio-chemically by-pass 'normal' reality and connect us to this inner world of light. How?

In 1952 it was discovered that there was a close similarity in the chemical composition of Mescaline (synthetic peyote cactus) and adrenalin. Further research revealed that lysergic acid (LSD) - an extremely potent hallucinogen derived from ergot - has a structural biochemical relationship to the others.

It was subsequently discovered that adrenochrome, which is a by-product of the decomposition of adrenaline, can produce several of the symptoms observed in mescaline intoxication. Adrenochrome also occurs spontaneously in the human body, which means that we should all be capable of manufacturing minute quantities of a chemical which is known to cause profound and far reaching changes in consciousness.

Do hallucinogenic plants merely stimulate something that already exists within our very being?

In his book 'Food of the Gods' Terence McKenna goes even further and suggests that psychoactive plants may have been linked with early religious belief, or even more profoundly, he attributes the development of spiritual consciousness itself directly to psychoactive substances and the subsequent mystical experiences they seemed so dramatically to evoke.

He writes 'If, as suggested earlier, hallucinogens operate in the natural environment as message-bearing molecules, exo-pheromones, then the relationship between primate and hallucinogenic plant signifies a transfer of information from one species to another. Where plant hallucinogens do not occur, cultural innovation occurs very slowly, if at all, but we have seen that in the presence of hallucinogens a culture is regularly introduced to ever more novel information, sensory input, and behaviour and thus is moved to higher and higher states of self-reflection. The shamans are the vanguard of this creative advance.'

He continues: 'The prolonged and repeated exposure to the psychedelic experience, the 'Wholly Other' rupture of the mundane plane caused by the hallucinogenic ritual ecstasy, acted steadily to dissolve that portion of the psyche which we moderns call the ego. Wherever and whenever the ego function began to form, it was akin to a blockage in the energy of the psyche. The use of psychedelic plants in a context of shamanic initiation dissolved, as it does today, the knotted structure of the ego into an undifferentiated feeling that Eastern philosophy calls the Tao.'

Millions of people all over the world who have experienced altered states of consciousness induced by psychoactive plants or drugs in the last thirty years also report similar feelings. There are reports of a heightened sense of awareness and sensory perception, an increase in the experience of cosmic love and beauty and a feeling of connectedness, or Oneness with all things. Time and space become distorted in general with heightened sensory perception, and there is an increase in the phenomena of paranormal experiences such as telepathy and out-of-body experience.

Some people who have had mystical, religious or spiritual experiences during altered states induced by plant or synthetic psychedelics, also report seeing everything as 'white light', or experiencing either cellular or cosmic phenomena usually invisible to the naked eye such as auras, magnetic force fields, primary cosmic rays and so on.

The psychoactive drugs do, indeed, seem to open hidden doors within the mind. In the words of William Blake, mystic, visionary, artist and writer - 'If the doors of perception were cleansed everything would appear to man as it is, infinite'.

Whatever biochemical reactions take place between plants and humans however, the desire to experience the mystical and spiritual phenomena appears to be an integral part of our collective humanity and our individual psyches.

In his book 'The Teachings of Don Juan: A Yaqui Way of Knowledge' writer Carlos Castaneda, reporting on his anthropological studies with Don Juan, a Yaqui indian who was well versed in the use of mind altering hallucinogenic plants writes: 'The importance of the plants was, for Don Juan their capacity to produce stages of peculiar perception in a human being. That he guided me into experiencing a sequence of these stages for the purpose of unfolding and validating his knowledge, I have called them 'states of non-ordinary reality', meaning unusual reality as opposed to the ordinary reality of everyday life. The distinction is based on the inherent meaning of the states of non-ordinary reality. In the context of Don Juan's knowledge they were considered as real, although their reality was differentiated from ordinary reality.'

The desire to know something of this inner reality, to know and to experience something of our own divinity is a real inner need that was once catered for in social ritual in our early nature religions. What was once considered a sacred and spiritual necessity, however, is now condemned to illegality the world over with few exceptions. The altered state induced by psychoactive plants is now considered to be of no use to society whatsoever, either individually or collectively.

Is it any wonder, therefore, that modern civilisation and society today, in general, is so bereft of an inner spirituality, plagued with psychological and physical sickness, ecological sickness, spiralling crime rates and pandemic negative socio-chemical drug abuse and addiction?

The emphasis in these early cultures was to obtain wisdom from the plants, and to enter an altered state (right brain). Although some ancient surviving rituals still exist in remote areas of the globe to this day, on the whole, we have lost our Shaman, Witch Doctors and our Medicine men, who performed the role of intermediaries between Heaven and Earth.

Although we have, in the main, lost our connection with these ancient spiritual practices, they can never be lost entirely and forever. Despite the suppression of these timeless rituals in Europe and America over the last few hundred years, in the last fifty years we have witnessed a phenomenal and almost instantaneous revival of these spiritual practices.

The use of LSD from the sixties onwards changed our society forever. A rich socio-spiritual culture rapidly evolved as a direct result, particularly affecting music and the arts, and the global social order diversified at the same time as the collective unconscious was undergoing a drastic evolution.

However, in the early sixties, the idea that LSD could cure ordinary psychosis initiated a period of intense medical research into the drug that was first synthesised in 1938 by the Swiss scientist Dr. Albert Hoffman. Conducted in both the UK and US, unsuspecting patients were given large doses of LSD, with the encouragement of psychologists such as R.D. Laing on the basis that it opened up the mind more quickly than normal therapy procedures. One doctor recommended his patient for LSD therapy because it 'cured all ills'.

Unfortunately the trial doses of the drug were far in excess of recreational doses, and in a series of experiments conducted by the CIA, unwitting patients were given LSD for 77 days running. The Misuse of Drugs Act in the UK, however, has now forced LSD - and the rest of the hallucinogens - firmly underground - although there is current talk of reintroducing medical trials with LSD in the UK as a cure for various mental health issues. Otherwise, it is paradoxically still a criminal offence to expand your mind in this way, despite its obvious and extremely long-standing relationship and contribution to human psychology, philosophy and culture.

Cannabis, a mild euphoric and psychoactive plant officially classified as a medicinal Herb, was once used as a cheap and ecological source of rope, paper and cloth, as well as being used for thousands of years in herbal remedies and religious rituals. This was said to be due to its ability to simultaneously relax and expand the mind.

Virtually unheard of in the West fifty years ago, an estimated 5 million people in the UK alone use Cannabis regularly for medicinal and/or recreational purposes. Some also believe that Cannabis has religious applications. In the

Rastafarian culture cannabis is seen as a sacred plant, a means of connecting to Jah (God), a way of stimulating the brain in a spiritual way. It is also used regularly as an aid to meditation and creativity in many homes throughout the world.

But does Cannabis really possess medicinal, creative or spiritual qualities?

Cannabis Sativa is listed in every herbal medical book - both ancient and modern - as a profound plant good for a large variety of medical and nervous disorders. Its attributes include the power to ease hysteria, insomnia, coughs, colds, women's pains (Queen Victoria used it for her PMT) and a variety of other medical conditions.

At the University of Tel Aviv new research data has now confirmed that a chemical present in Cannabis is also naturally present in the human brain itself! Is Cannabis a natural brain food?

The latest research conducted by the Israeli scientist Dr. Maschon Knoller, head of neurosurgery at the Tel Hashomer Hospital in Tel Aviv, has proved that a new drug extracted from the cannabis plant could literally save the lives of brain injury victims and recent research confirmed that the active ingredient in cannabis has the actual ability to stop brain cells killing themselves after serious head injury trauma, which normally leads to paralysis and in the worst case scenario, to death.

Scientists have been searching for a way to prevent the damage in head injuries caused by a chain reaction that leads to the destruction of irreplaceable brain cells. The answer seems to be in the Cannabis plant. Could Cannabis in fact be good for our brain cells in general - improving our psychological state?

As far as the medical application of Cannabis is concerned Asaf Yefet was one of the first trial patients to be saved by a newly formulated cannabis extract drug called Dexanabinol that claims to prevent serious neurological handicaps after serious brain injury.

Having fallen from his horse Asaf suffered severe head trauma and was taken to the Hashomer Hospital, where he was diagnosed as having a 3 percent chance of survival. He was instantly put on Dr Knollerís trial and within hours he was being given Dexanabinol through a drip in an attempt to heal the damage to his brain. (The drug must be administered within two hours of the original injury.) After just five days Asaf regained consciousness and was able to walk home from hospital just three weeks later. It took him another three months to make a full recovery - totally against all the odds!

The active ingredient in cannabis seems to inhibit the production of glutamate, the scavenging enzyme usually apparent in small quantities in the brain, which is responsible for the effects of brain damage. Professor Aviv who also works on the programme believed that this plant could provide the main ingredient for a miraculous cure for treating brain injuries, haemorrhages and strokes.

Bill Alker of the National Head Injury Association says Dexanabinol, extracted from the cannabis plant, could have a 'major impact' saving lives and preventing permanent disability. A third of all brain injury victims currently die whilst the rest are often confined to wheelchairs. Could this type of cure be useful in Meningitis also? Cannabis has certainly proved effective for Multiple Sclerosis sufferers, so much so that many have now turned to it for regular pain relief.

Sir Thomas Culpepper, the famous 16th century herbalist, referred to the many qualities of Cannabis, which he refers to as Hemp. He tells us that it is good inflammation of the head, or 'any other parts'. It eases the pain of gout, knots in the joints, sinews, expels wind, as is good for hot or dry coughs. The plant is also good for jaundice, eases colic and prevents nosebleeds. Cannabis is also good for glaucoma, insomnia and nervous disorders.

Cannabis also has the advantage of being mildly euphoric and psychoactive - no doubt because of its relaxing effect on the brain and nerves, perhaps enabling us to induce alpha waves - and for this reason perhaps it is so currently popular.

Do people smoke cannabis because they are naturally, instinctively and biologically drawn to a plant which helps stimulate and protect their own brain function in an extremely stressful world? Does smoking cannabis help relax the mind and help induce more relaxing and spiritual alpha waves? Does smoking Cannabis help switch off the intellectual left-brain, and activate the imaginative right-brain?

Although until now society in general may consider the altered state and the use of psychoactive plants of no social or cultural validity whatsoever, I think, on the contrary, that these plants have much to offer in medicinal, psychological and spiritual practices.

Are we seeing a spontaneous revival of an ancient form of tribal and Shamanic ritual that has now become rapidly incorporated into our modern culture? Since many of the great artists, musicians, writers and inventors of recent years - who have contributed so much to our own modern culture - have all, at their own admission, participated in mind altering and expanding substances of one sort or another.

According to recent polls, more people take recreational psycho-actives in Britain than in any other country in Europe. Is it a coincidence that the UK has the largest use of psychoactive drugs, and at the same time has one of the most creative, artistic and successful musical industries, not just in Europe, but in the World? Can this unique creativity be attributed in part to the use of plant psychoactives.

Consumed daily by millions of people worldwide, despite every effort and attempt by the authorities - the 'status quo' - to stop them, many people are now consciously choosing to alter their consciousness with plant psychoactives on such a regular basis, that it is proving to be a hopeless and impossible task to prevent them. Of course it is! How on earth do you police billions of people all round the world every day who want to relax, alter their brain waves and enter a more spiritual and imaginative state of consciousness? And more importantly, does society have a moral right to do so?

In our primal societies however - which did not suffer the crime and disorder that we do - the regular ritualistic ingestion of naturally grown psychoactive mind expanding potions kept the individual and society balanced. (Left brain/right brain).

Perhaps, instead of trying to suppress what has been considered sacred and spiritual for so long, we should embrace the altered state and perhaps we would see a reduction in what, today, has become a glaring social and spiritual malaise, and would hopefully see an increase in the number of happy, whole and well adjusted individuals.

Many still unconsciously seek this inner wholeness and spiritual experience of inner religious ecstasy through the psychedelic experience, but whereas the seeker of spiritual wisdom would once have been guided through his profound and intense experience by a 'shaman', one who knew of the spirit worlds who could guide the seeker through the labyrinth, the seeker today is forced into a criminal underworld and sub-culture where mind expanding drugs are often sold alongside deadly mind numbing narcotics - such as heroin and crack-cocaine - and where they are left to explore these higher realities of consciousness alone.

The altered state and the psychedelic experience are, by their very nature, intense and powerful, and should not be undertaken lightly. As in all walks of life, however, there are always casualties and the line between illumination and psychosis a thin one.

As far as Cannabis is concerned, due to the importation restrictions on the natural strains to the UK people have now turned to mass production of artificial home-grown strains known as 'skunk' to satisfy a growing demand, and a lucrative gap in the market. However, the unfortunate side effects of the chemical imbalances in artificially grown plants can cause serious mental health problems in susceptible individuals. Prohibition merely drives the problem underground and can have unforeseen consequences that merely make the problem worse.

Whatever the risks, however, there are always those who are prepared to take them in search of the altered state. This drive to explore the boundaries of inner space is, with the right guidance, no more or less dangerous than the desire to expand the boundaries of outer space. Both are willing pioneers and explorers who know the risks and who take them in order to expand our understanding, whether it be of the inner or outer worlds.

The current political attempt at prohibition of mind expanding plants and drugs will never work because humanity has always sought to expand its individual and collective consciousness since the beginning of culture as we know it.

Whilst tobacco and alcohol remain the biggest drug killers in society today, and yet remain legal, the psychoactive plants that seem to evoke profound creativity and spiritual development remain totally illegal. The desire for some form of spiritual 'ecstasy', however, is embedded in our genetic and psycho-spiritual make up. The memory of our collective tribal past and our relationship with nature, the knowledge that within the self is a state of being which transcends normal waking consciousness is etched eternally on our own genetic make up. Our individual DNA carries the collective memory of human history - and it is pumping around in our veins to this day.

Although Aldous Huxley was convinced that the key to solving the world's problems lay in changing the individual through mystical enlightenment and the psychedelic experience, I think it will be some time yet before the rest of the world catches on.

Sound, Music and Sacred Dance.

'Music is a moral law. It gives a soul to the universe, wings to the mind, flight to the imagination, a charm to sadness, gaiety and life to everything. It is the essence of order, and leads to all that is good, just and beautiful, of which it is the invisible, but nevertheless dazzling, passionate and eternal form'.

Plato

Because everything is energy, or vibrations, music has been a powerful tool in spiritual and religious ritual for altering states of consciousness since our first nature religions evolved, and to this day, in every religion worldwide, including the monotheistic religions, Judaism, Christianity and Islam, singing or chanting, and the playing of sacred music has a primary importance in religious celebration and ritual.

Specific rhythms, patterns, sounds and tones can have magical effects on the listener, as they transport you to higher levels of consciousness. You know that yourself. Think of your favourite piece of music, and how it transports you into a higher realm. You forget all your troubles as you are captured by pure sound. Certain notes make the back of your neck tingle, or make your hair stand on end. Music is a powerful emotional release. Interestingly, humans and animals share similar responses to music and the 'chill factor'- when we get shivers down our spines listening to certain notes - can also be observed amongst dogs when suitably emotional music is played. This was recently proved in a study by Harriet Read, a postgraduate from the University of Sheffield in England.

Music, or sound, as we discovered earlier, has the ability to affect the shape of material objects, and further research into music has found that different areas of the brain light up during a 'musical experience'. We know that it is the right brain that processes music, so perhaps it is the right brain (feminine, intuitive, mystical) that becomes activated and stimulated by sound, enabling the listener to enter higher states. Chanting has the double effect of altering brain activity and vibration of the chanter, as well as focusing the mind in contemplation.Today, music is being used in medical therapy with remarkable results in such disorders as autism, where music becomes a language that the child can understand, and by which an avenue of communication can be established between the patient and the therapist. It has also been found that music is, in effect, a vital life force, which we all need for our well-being. As sound and vibration have a large role in universal creation, differing sounds have a large role in the state of consciousness.

Music is a universal language which we all understand and which we all need. Music (sound) is a life force in its own right. Music, or individual specific sounds can also act as an aid to meditation and the experiencing of mystical states of consciousness similar to the ones we have already described.

To the Bushmen of the Kalahari the trance dance is a vital part of religious ceremony. The Bushmen believe that when they enter a trance they become imbued with potency and enter a spiritual realm where they can contact God, gain healing powers, and see visions. According to the Bushmen, the trance dance is older than history and has in it the seeds of all religious philosophy.

That music, which is a cohesive organisation of sound into a meaningful and distinct melody, can deeply affect our own vibrations through our magnetic fields, affecting our thoughts and feelings, has long been incorporated into religious and spiritual worship, and no religion of whatever origin or culture, has omitted to use music within the celebratory or ritualistic ceremonies. In some cases, particularly in early tribal society, the trance state was seen as highly desirable and frequently easily obtainable.

For thousands of years our ancestors have used the altered state as a means of finding spiritual guidance from higher spiritual or mystical sources, and most, if not all, religious or spiritual systems are based on the insight gained through the vision, revelation or divine mystical experience.

Mysticism is the one true source of our religious beliefs and I believe we all have the right to that experience. The mystical experience is not just for holy men and prophets, it is for us all, if we should care to seek it - or it should seek us - and if we are psychologically prepared and ready for it.

The altered state can also be achieved by inducing a mild hypnotic or trance state, by the use of ceremonial and sacred sound and rhythm. This is observed in every religious ritual on the planet, and all of them use sound as a means to alter states of consciousness suitable for spiritual worship.

However different the musical culture the effects of sound remain universally the same. The repetitive and hypnotic rhythms and movements, the note, the tone, the speed, the intonation - each affects the feelings and mood of the player and the participants. Music and sound acts as a focus for meditation. Since music does not involve an intellectual response (left brain), since it stimulates and requires more imaginative responses, it automatically activates the right brain.

Altered states are frequent in prolonged dance as the conscious mind becomes quietly tranced and hypnotised by the rhythm and the brain waves alter to Alpha.

Dance student Naomi Friedlander says that to her, dance represents a meditation and that often, she feels as though she were detached from her body, as in the out-of-body experience. To her dance is something essential, something that expresses reality at a deeper soul level. Dance, says Naomi, is the freedom to express emotions, ideas and feelings through an abstract art form that has the potential for individual human ecstasy and union with the divine within its sacred rhythms. Naomi believes that these rhythms are within us, as well as without, and that by using music as a meditation, we can elevate our own state of consciousness or awareness.

The Dervish dancers of Turkey are known for their intricate dances that are based on cyclic movements that are said to represent the cyclic movements of the heavenly cycles. The Dervishes twirl and whirl until they are close to a semi-hypnotic trance, whereupon a blissful mood sets in, and the dancer becomes open to the realms of his imagination and intuition, through the stimulation of the right brain. Some dancers even claim that the physical world appears to disappear into white light after hours of exertion.

Even more surprisingly, in a recent paper published by Andriano Lameira and Marcus Perrimen entitled 'Great Apes Reach Momentary Altered States by Spinning', (2023) the authors note that like humans, great apes spin at speeds that induce physiological highs, sometimes up to 100 times before stopping. 'When spinning, the apes achieved sufficient speed to alter self perception that were comparable to those transcendental experiences in humans. Thus spinning serves as a self-sufficient means of changing body-mind-spirit awareness in both apes and humans.'

Spinning induced highs in apes also suggested that, like humans, great apes voluntarily sought to engage in altered experiences of self perception and situational awareness. 'We are obviously not alone in the intentionally induced altered states of consciousness. Archaeology, history and ethnography show that these mind altering rituals have taken place since the beginning of civilization.'

Chanting and meditation are also means to induce altered states and are quintessential to Buddhism. Certain phrases or keywords are repeated over and over to establish a vibration, or sound which will elevate and alter the state of consciousness of the participant. Repetition of certain sounds, words and rhythms has a hypnotic effect on the mind, enabling increased right brain function and altered brain waves.

Sufi texts record the optimum sounds for enlightenment in relation to chanting, and these are the three basic long vowel sounds of A, I and U. These are what the Sufis call the 'Universal Harmonic Constants' and they are used in all the mystical paths that utilise sound. 'Master these sounds and you will ascend the stairway to heaven'. (16th Century manuscript).

The Tibetans have been accessing altered states with harmonics for thousands of years and the Indian yogis have used sacred sound for millennia. The Aborigines of Australia say they travel to their Dreaming on the waves of harmonics. In Chinese music the instruments are said to symbolise and emulate the sound of the elements, to reflect the sounds of nature. The Flute represents the wind in the trees, the drums symbolise the heartbeat, or thunder, and the use of bells represents the flow of water. By imitating natural sounds and rhythms, the listener can be induced into an altered state, transported through different vibrations to different levels of being and consciousness.

The 20th Century Mystic and philosopher Rudolf Steiner also observed that a particular use of tone opened perception to well beyond the current version of reality. The altered state renders the mind receptive and sensitive to higher invisible vibrations and energies. Self-hypnotic trance phenomena including that induced by music, can precipitate personal and psychological experiences, ranging from feelings of bliss, transcendence, immunity to pain, contact with the world of spirits, which can include associated feelings of universal love and oneness with oneself and with all of nature and all of creation. Visions are also common to the altered state.

As we have already observed, brain rhythms have a lot to do with what level of mind activity we are engaged in. Altering brain waves helped to enhance and evoke the visions and dreams normally only experienced at lower brain wave frequency that is usually prominent in the sleep, dream, or meditative state.

These rituals were employed for well over ten thousand years in our collective history but the advent of monotheism, Western materialistic technology and science, brought an end to these 'pagan' rituals and practices, and the fact that Western society is now suffering severe social problems, is no doubt a symptom of its own collective denial of the mystical feminine right brain function, and its greater respect for male dominated left brain intellectual and materialistic achievement.

Although these rituals and traditions have largely died out as tribal peoples are assimilated into modern culture, there would appear to be a massive revival at the instinctive level, in society as a whole, to reinstate these rituals and to redress the imbalances of the psyche.

Meditation, Trance and Transcendental Experience.

'You can't see it, because it has no form.
You can't hear it, because it makes no noise.
You can't touch it, because it has no substance.
It cannot be known in these ways, because it is
All embracing Oneness.
It is not high and light, or low and dark.
Indefinable yet continually present,
It is nothing at all.
It is the formless form, the imageless image.
It can't be grasped by the imagination.
It has no beginning and no end.
This is the essence of Tao.
Stay in harmony with this ancient presence,
and you will know the fullness of each present moment'.
                              Lao Tzu - 5th Century BC.

Meditation as we know it originated in the East. Meditation is a means of entering into altered states of consciousness by non-action.

In meditation, the body and mind are subdued, not by vigorous movement as in trance dance, but by non-movement.

In meditation, one seeks illumination by contemplation. The goal of meditation is to find one's own inner peace, one's own Nirvana. Meditation requires stillness, quietness, calmness, breathing techniques, and shutting the eyes.

Why do we have to close our eyes? Quite simply, producing alpha waves is physiologically linked to the eyes. Closing the eyes immediately induces the brain to produce alpha waves that are essential for the trance state. Furthermore, turning the eyes upward, to the middle of the forehead, the seat of the pineal gland, or Third Eye centre, and the alpha wave is induced even faster. This is why we shut our eyes.

In The Zohar, the book of Jewish Law it says quite clearly:

'The secret is the eye. Come and see. The secret is: close your eyes and roll your eyeballs. Those colours that shine and glow will be revealed. Permission to see is granted only with eyes concealed, for they are high and concealed, overseeing those colours that can be seen but do not glow.'

It was by these same meditation techniques that Buddha found Nirvana, within himself. The concept of inner enlightenment is also inherent in Christianity, and Christ's words 'The Kingdom of Heaven lies within, seek and ye shall find, knock and the door shall open.'(Mathew 7.70, is a direct reference to the potential for inner spiritual enlightenment.

Christ also says 'Let thine sight be single (a reference to the third eye) and thine whole body shall be full of light.' We already know that our own bodies, and the bodies of all living beings are suffused with an inner light - is this what Christ meant by this inner kingdom of heaven, this inner kingdom of Light? Can we all experience nirvana, can any of us tap into this inner domain?

In 1994 Dr. Michael Persinger of Laurentian University in Canada found that he could stimulate a profound mystical experience in all of his volunteers, whatever their religious persuasion, by stimulating the Temporal Lobe with magnetic forces. These experiences ranged from extra sensory perception, to experiencing 'God' and 'heaven' and the 'oneness of things' '.

The mechanism he used to stimulate these experiences is simply called - 'The God Machine' and the experiences Dr. Persinger's volunteers were remarkably similar to those reported by people who have ingested hallucinogenic plant material or who have practised long periods of meditation. The theory that hallucinogenic plant material and meditation may also be stimulating the temporal lobe in a biochemical way should not be discounted.

From Dr. Persinger's research, we might conclude that each of us contains the potential for divine, mystical experience. It merely needs the relevant stimulation.

In 'Autobiography of a Yogi', Paramahansa Yogananda wrote of his own 'Heavenly' experiences stating: 'Sri Ykteswar (his teacher) seldom indulged in riddles. He struck gently on my chest above the heart. My body became immovably rooted, breath was drawn out of my lungs as if by some huge magnet! Soul and mind instantly lost their physical bondage and streamed out like a fluid piercing light from my every pore.

The flesh was as though dead: yet in my intense awareness I knew that never before had I fully been alive. My sense of identity was no longer narrowly confined to a body but embraced the circumambient atoms.'

He continues 'All objects within my panoramic gaze trembled and vibrated like quick motion pictures. My body, Masters, the pillared courtyard, the furniture and floor, the trees and sunshine, occasionally became violently agitated, until all melted into a luminescent sea; even as sugar crystals, thrown into a glass of water dissolve after being shaken.

The unifying light alternated with materialisations of form, the metamorphoses revealing the law of cause and effect in creation. An oceanic joy broke upon the calm endless shores of my soul. The spirit of God, I realised, is exhaustless Bliss; His body is countless tissues of Light.'

Finally, he stated that; 'The divine dispersion of rays poured from an Eternal Source, blazing into galaxies, transfigured with ineffable auras. Again and again I saw the creative beams condense into constellations, then resolve into sheets of transparent flame.

By rhythmic reversion, sextillion worlds passed into diaphanous lustre, then fire became firmament.'

The doors of perception exist within each of us, and I believe that we all have an inner bio-spiritual necessity for these higher states of consciousness. After all, this greater reality that is perceived in the altered state, is intrinsically part of who and what we are, and to deny it is to deny half of ourselves.

We are spirit manifesting as matter, and the mystical revelation is an experience sought after and needed if we are to acknowledge our own divinity and our own need to experience that divinity.

Our divinity is the part of ourselves that is loving, just and moral, it is the part of ourselves which is more than mere mortal, it is the soul, the part of ourselves that is divine in its truest sense.

Indeed, it is surely our spiritual awakening and understanding that holds the key to our higher evolution as a species, for in it, is the higher being of mankind, and the hope for our future and the future of this world.

This inner world, this inner reality, has long been sought after by mankind, and the vestiges of truth which remain in the world's religions which should enable us to connect with this higher reality and which are evolved to bring the participant within reach of his inner kingdom, his inner heaven, have been replaced by meaningless outer form and ritual.

As Jung so aptly pointed out, unless the experience of God is one that is experienced within our own souls, the external form of worship will amount to absolutely nothing.

Miracles and Magic.

'Thought, feeling and the spoken word are the only creative forces in the Universe.'
Saint Germaine.

The eminent 20th Century Cambridge philosopher, Dr. C. D. Broad once wrote, 'We should do well to consider much more seriously than we have hitherto been inclined to do, the type of theory which Bergson put forward in connection with memory and sense perception. The suggestion is that the function of the brain and nervous system and sense organs is, in the main, eliminative and not productive. Each person is at each moment capable of remembering all that has ever happened to him (or her) and of perceiving everything that is happening everywhere in the universe. The function of the brain and nervous system is to protect us from being overwhelmed and confused by this mass of largely useless and irrelevant knowledge, by shutting out most of what we would otherwise perceive or remember at any moment, and leaving only that very small and special selection which is likely to be practically useful'.

In other words, the conscious mind tends to focus itself on the business of physical survival, filtering out all the other data that we are constantly receiving from the universe via the electromagnetic force field, or aura. The altered state, induced either by a temporary chemical input, or moments of contemplation and meditation, allows us to glimpse this inner world of energy for a short time only. If we were to stay any longer in these states of blissful consciousness, we would probably become incapable of functioning at the material level.

However, the all too brief glimpses we have of this higher world of energy and geometric patterns, light and lucidity, have at least confirmed the existence and validity of higher states of consciousness. In these states of higher consciousness, we find that the power of the mind is unlimited and if we were able to maintain this connection and focus our mental power, we would find ourselves capable of performing so-called miracles and magical feats. With an accomplished awareness and belief in mind power alone, the normal is transcended, and with the total use of mind power miracles and magic are not only possible, but also actualized.

Great devotion and study is required to comprehend the laws of nature, and by accessing the unlimited power within the mind, within our own consciousness, anything is possible. Magic and miracles are, if we could truly understand them, be seen to be not so much magic or miracle, but part of the normal fabric of the mystical universe, and fully possible to those who develop their inner power and find the connection between themselves and the forces of nature.

Mind power is unlimited. It is only the conscious mind that limits the idea of power by its own disbelief. The power of the mind to transcend matter is not just a vague concept, or the fanciful rantings of the logically challenged, it is a real possibility and there are often simple explanations.

The Bible itself is full of stories of magic and the miraculous, and what more miraculous feat could there have been in the New Testament, than Christ walking on water. Now, straight away, you have put this idea into the category of 'impossible' in your own mind, and yes, if you tried to do it now, you would be unable. You would lack faith.

And yet today, scientists are working on simple anti-gravity mechanisms, such as the ceramic superconductors used to propel tons of train metal effortlessly down the line whilst levitating six inches off the ground. In Finland, scientists are about to reveal details of what may be the world's first anti-gravity device, which significantly reduces the weight of anything suspended over it. If harnessed, the antigravity effect could be turned into a means of flight. The researchers, from the Tampere University of Technology in Finland believe this could form the basis of a new power source that would have endless beneficial uses.

Were similar anti-gravity devices available to the Egyptians and Celts when they built the gigantic edifices of the Pyramids and Stonehenge? Are the stories of Merlin, who was said to have transported stones magically by levitation from hundreds of miles away, now looking more like fact than fable? No modern engineer, mathematician, or builder has yet unravelled the mysteries of these two great edifices. Did the Egyptians have an arcane knowledge of anti-gravity when they built their Pyramids, lifting massive stones effortlessly into the sky?

And what about all the people, who have appeared, or claimed to be able to levitate? Is this really possible in any shape or form? We know that our bodies are made of magnetic forces. Is it so impossible to believe that a highly evolved being could bring his own magnetic field into a specific alignment, and use them to levitate? This, in effect, is what Christ did as he walked over the water, being adept at mind over matter, he was easily able to transfer his total consciousness to his magnetic field, and, diminishing the reality of his physical body, converted his focus to his spiritual body (which weighs less than one quarter of an ounce) - and was thus enabled to literally walk on water.

Of course, miracles of this sort are out of reach of the majority of mankind, but there are humans on this planet today who have developed their mind power to such a degree, such as the Yogis of India, that they are enabled to perform miraculous feats.

Miracles, however, are merely the total application of the laws of the universe by the individual, who has developed and identified with his inner power to such a degree that he (or she) has become one with it, and master of mind over matter.

The study and practice of Magic represents the practice of certain rituals, usually involving the use of sacred symbols, words and invocations, which are sometimes employed to help focus the mind to achieving these higher states of consciousness. What we call 'miracles' and 'magic' are, in essence, produced by operating on a much higher level of consciousness (vibration) than that we use to survive in our everyday world.

Although we are not ready as a species for such mastery at this stage of our evolution, we are progressing and evolving at such a rate that it may not be long until we are all able to perform such supernatural feats ourselves as a matter of course.

Mind power alone is responsible for all of creation, mind power alone is responsible for our thoughts and actions. If you believe you are capable of something, then you will be, if you doubt, then you will fail. By connecting to the divine in oneself, anything is possible. By connecting to the divine in oneself, one can truly have mastery of mind over matter.

The Power of Positive Thinking.

'Every moment of your life is infinitely creative and the universe is endlessly bountiful. Just put forth a clear enough request, and everything your heart desires must come to you'.

'Creative Visualization.' Shakti Gawain.

We can all bring a little magic into our own lives, without recourse to elaborate magical rituals. This is done by the positive use of thought - positive visualisation - the controlled use of mind power. And its principles are really quite simple.

Thought becomes form, therefore change the thought, you change the form. Negative thoughts create negative actions and positive thoughts create positive actions - and reactions. If you convince yourself you will fail, you will fail. If you convince yourself you can succeed, you will succeed. Optimism is contagious!

Michael Flately, who wrote and choreographed the inspiring dance spectacles Riverdance, and Lord of the Dance, came from a poor Irish family. Although Michael would spend most of his time at school staring out of the window daydreaming, much to the annoyance of his teachers, he claims he was busy visualising his future! 'I can't emphasise it enough' he said 'just keep visualising what you want in glorious colour and it will come to you'.

Bill Gates, founder of the Microsoft Corporation visualised a computer in every home as a dream of things to come - when computers were still main-frame! His dream came true. He just kept on visualising!

How much does will-power and positive visualisation of one's own destiny affect that destiny itself? Can positive thinking really help us improve our lives and achieve extraordinary feats?

There are many new age self-help books now available that encourage us to 'think positive thoughts' and 'visualise' what we want for ourselves. We are told we can transform our lives in this way. Can thought really work in such a profound way?

In recent experiments conducted in a US hospital, positive thoughts and prayers were said for half of an experimental group of patients, whilst the others had no prayers said for them at all. In the meticulously controlled experiments, it was found that the group of patients who had been prayed for recovered quicker by far than the group who had not been prayed for. None of the patients were aware of which group they were in.

In his book 'The God Experiment' author Russell

Stannard documents the experiments conducted at the Mind/Body Medical Institute, New England Deaconess Hospital, Boston, Massachusetts, which is run by the project's chief investigator Dr Herbert Benson. Dr Benson found crucial evidence to support the power of prayer, which was substantiated by a study performed by R.C. Bird who also found similar test results.

Prayer is the power of thought intensified and directed at a higher mind source than the individual self. People who pray in groups intensify the mind power considerably. Prayer, meditation or contemplation is used in nearly all religions as a medium to higher states of consciousness.

The power of prayer is extremely potent, of course, praying is not just about asking for material things, we can help people, protect people and help heal people with our positive focused thoughts. Lack of thought or care for people also have equally powerful effects. The official prayers of blessing and protection that were said for Princess Diana were stopped very shortly before her tragic death.

The prayer and blessing normally considered essential for the launching of new ships 'God bless this ship and all who sail in her' - were omitted on all of the following -The Titanic, The Britannic and the Olympic. White Star lines did not believe in such time-wasting formalities. Two of these giant vessels, The Britannic and the Titanic, both met tragic and dramatic fates. The Titanic sank on her maiden voyage to New York after colliding with an iceberg with the loss of over one thousand two hundred souls, the third, the Olympic, was involved in a dramatic mid-sea collision. The prayers that were normally said for all newly launched ships were omitted when these three giants set out to sea for the first, and in the case of the Titanic, the last time. Is it a coincidence that none of them were blessed, and that all three of them met deadly fates?

Can our thoughts affect the outcome of events? Recent research in the US has proved conclusively that in experiments where the volunteers had to will various numbers up on a random number generating computer programme, the volunteers themselves were distinctly able to influence the outcome of what was apparently random number selection, by willpower alone.

Dr. Lev Pyatnitsky, a physicist at the Russian Academy of Sciences, has been studying the ability of people's minds to affect the environment. Using tap water Dr. Pyatnitsky proved in a study of l5 volunteers that six were able to focus their minds on the water sufficiently to cluster the molecules more tightly together. 'Statistically' he said 'it is like tossing a coin and getting heads billions and billions of times.'

As we have already seen, the primary agitating factor on the pre-physical body is thought. Thoughts, moods and feelings create electromagnetic changes in the auric field, which eventually crystallises onto the material plane.

Now, a thought and an action is not just something that affects you, whatever you think and do will have a corresponding effect on the rest of the universe. The vibrations that you set up for yourself, which are a direct result of your own thinking patterns, will eventually and absolutely result in the resonance within the universe which will attract like vibrations.

For example, if I hit a tuning fork on a suitable surface - let's say for the sake of argument tuned to the note of C - every other object in the vicinity which is tuned to the same C will start to vibrate in accordance. So, our thoughts are like the tuning fork, whatever frequency we are vibrating at, all the other vibrations of the same frequency will start to vibrate with us. Like attracts like. However you are vibrating, you will activate others vibrating at the same frequency. Therefore, by improving our own frequency, our own vibration, we will attract that which vibrates in accord.

The more we believe in, and focus our mental power on the outcome of our efforts in a positive way, the more likely they are to happen. This is now scientifically proven.

The vibrations you set up for yourself through your thought habits and patterns will, believe it or not, eventually manifest onto the material plane from the higher energy which is the source of thought, and you will then find that the world you inhabit, is a world of your own creation. We can change it, improve it for the better, by using our powers of visualisation, (right brain), for, as we saw in the examples on hypnosis, the mind is very sensitive to suggestion

It has recently been proved that activated nerve cells in the brain use oxygen at a faster rate than inactive ones, and so the draining of oxygen in a region indicates the presence of activity in the other. Repeated stimulation of one particular area actually creates new and more ingrained neural pathways. This is obvious when you think of any new task you have set yourself, which proves difficult and slow to begin with, playing the piano, for instance, but after constant practice and repetition, becomes second nature.

Mental visualisation actually changes the shape of the brain! In his book 'Mind Sculpture, Your Brain's Untapped Potential' author and researcher Ian Robertson has proved that by repeating certain tasks over and over again, the brain actually creates new neural pathways for itself, and once the task has been repeated sufficiently, it actually becomes part of the brains permanent neural network. He writes: 'Can we actually see what happens in the brain during mental practice? Fortunately we can. Using a PET scanner, a brain scanner that shows which parts of the brain are active during different mental and physical tasks, people were studied as they imagined themselves moving.

The movements were based on a simple task - manipulating a joystick, and the brains were watched to see which areas were switched on when people imagined moving it. The results were compared with what happened when the same people simply got ready to move the joystick without actually moving it.' He continues: 'It was also discovered that very similar parts of the brain lit up in these two different situations. In other words, mentally imagining a movement triggers much the same brain machinery as does preparing to make the same movement. It seems therefore, that imagining a movement is not very different from actually making the same movement as far as the brain is concerned.'

The act of positive visualisation actually affects the brain in a positive way. Imagine if you kept imagining it, and set up positive vibrations for yourself, and then it materialised for you!

Positive thinking is a means of creating new opportunities for yourself. The power of positive thinking, the clear, controlled and creative use of imagination represents the ability of the individual to harmonise with, and access universal energy. As we have already seen, there are higher, more subtle energies that exist within the individual, which exist on increasingly finer levels of energy and vibration - ultimately pure universal energy. You can tap into it.

These higher levels of energy, originally manifested from pure thought, are extremely sensitive to our thought vibrations and will actually and absolutely manifest as a virtual reality the physical manifestations of those thoughts - whatever they may be. This is echoed in the Biblical quotation: 'As a man thinketh in his heart, so shall he be'. (Proverbs 23:7)

As we saw earlier, changes are firstly produced on the auric field (electromagnetic force field) before they manifest onto the physical dimension. By understanding this simple process, we can try to control and direct our thoughts through positive visualisation.

Positive thinking actually enables us to create new and more harmonious patterns for ourselves in our lives, as we consciously make use of the flexibility and impressionability of the unconscious mind in our attempt to harness the power of mind over matter. We just have to tap into it.

Chapter Four.
FUTURE MIND.

Space-Time and the Sixth Sense.

'We can never fully know; I simply believe that some part of the human self or soul is not subject to the laws of space and time.'

Carl Jung.

Before the Big Bang and the creation of the materialised universe, there was no time and no space. Time and space only came into being with the birth of the physical universe. Can you imagine a time without time? Space without space? A time and space where time and space simply did not exist?

Although scientists once thought of time as being absolute, that is, fixed and inflexible, Einstein proved conclusively that time was not absolute, but was absolutely relative. Time, said Einstein, was a phenomenon that we perceived relative to the speed of light and the position and velocity of the observer.

Now, as we saw earlier, when we look into deepest space, we are in effect looking back in time, due to the fact that light takes so long to travel from one place to another. Conversely, if any conscious beings who should inhabit a distant planet located in one of those regions were to look into deepest space and see us here on Earth, they would only see the point in Earth's history relative to their own distance from Earth in terms of light years - that is, depending on how long it took the light from Earth to reach them.

To make this point clearer, let us imagine hypothetically, that on Planet X, which is many light years away, they have just picked up the images that were emitted when Cleopatra was on the throne of Egypt. Planet Y, however, a bit further back is just picking up light waves emitted when the first apes began standing upright! On Planet Z, however, further away than both of them, they have only just received the images, or light waves, emitted during the Jurassic period, and all they would see on Planet Earth would be dinosaurs! If each were to discuss the 'current' events on Earth with each other on their inter-galactic mobiles, each would have an entirely different view of Earth's time in history. All, however, would be accurate.

Time is not absolute, and time and space are both relative concepts given the position and velocity of the observer. Our perception of both is entirely relative to our own viewpoint in the apparent field of space-time as perceived by the five senses. Time too, takes on a relative quality in our individual lives. You have no doubt experienced that for yourself in your own relationship with time.

Some days - the days when you're happy and having fun - don't they seem to absolutely fly by - whilst others, those when you are sad, lonely or depressed, seem to drag on forever? Neither does this variable perception of time appear to be a logical or intellectual variance of perception. Rather, it appears to depend almost entirely on the emotional frame of reference of the individual concerned. In essence, however, how we generally view the universe and how we perceive the phenomenon of space-time is programmed into our biology so that we consciously only perceive a small segment of the total spectrum of sound and light, so that we perceive space-time and the universe in one particular way. The way we are currently familiar with.

What appears to us to be a logical and spatial sequence of events in a space-time continuum, however, is, in reality, only a view from our own point of view. In essence, we generally only perceive the events that are occurring at the vibration at which the material universe is operating in the same time-space vicinity as ourselves.

If we did not filter down the totality of sensory input, which is constantly bombarding us from the entire universe, we would be totally overwhelmed by it. We are selectively programmed to perceive only a small fraction of the electromagnetic spectrum, which determines our perception of the material dimension. Our consciousnesses have evolved to narrow down, filter down, all the cosmic data to help us survive biologically in the material world.

Occasionally however, as we saw in the previous chapter, we are sometimes able - sometimes permitted - to see behind the veil of 'normal' perception and experience a glimpse of different modes of awareness, or witness different realities, including different frames of reference to the time/space experience we are familiar with.

The altered state nearly always distorts the perception of time and since we were originally created from an invisible pre-space and time energy, could we not perhaps contain the essence of this pre-time and space energy within us, at the very core of our sub, or even pre-atomic being?

In 'The Tao of Physics' Fitzjof Capra writes 'The Eastern sages too, talk about an extension of their experience of the world in higher states of consciousness, and they affirm that these states involve a radically different experience of space-time. They emphasise not only that they go beyond ordinary three-dimensional space in meditation, but also - and even more forcefully - that the ordinary awareness of time is transcended. Instead of a linear succession of instants, they experience - so they say - an infinite, timeless, and yet dynamic present.'

In modern quantum field cosmology, physicists now agree that in essence, there is no time and that the past, present and future are one all encompassing reality. The situation, however, is quite different in the general theory of relativity. Space and time are now seen as dynamic quantities, or qualities, and that when a body moves, or a force acts, it affects the curvature of space and time - and in turn, the structure of space time affects the way in which bodies move and forces act! Space and time not only affect, but also are affected by everything that happens in the universe. Even more remarkably, the mathematical field theory suggests that positrons can appear to be positrons moving forwards in time, or as electrons moving backwards in time!

At the University of Nevada, Dean Radin has conducted experiments that suggest that volunteers were able to distinguish emotionally, and react to the difference between sad or happy images generated at random by a computer, five seconds before the image itself was transmitted!

Can the human mind travel both backwards and forwards in time? Can we perceive these movements of positrons and electrons backwards and forwards in space-time at an intangible, at an unconscious level, at a subatomic level?

The mystical and quantum theories of time would both seem to suggest that, somehow, time is an illusion, and that in reality 'time' does not exist at all. Furthermore, all events that we perceive to be happening in space-time are all actually happening at the same time!

Before 1915, however, space and time were thought of as being fixed - a dimension in which events occurred, but which was not affected by what happened in it. This was true even of the special theory of relativity. Bodies moved, forces attracted and repelled, but time and space simply continued, unaffected. It was natural to think that space and time went on forever.

How we perceive space-time with the limitations of our five senses make it impossible to understand this phenomena in any logical way. We can only do this in some form of altered state of consciousness. Time and space, we can only say, are relative to the observer. If then, time and space are relative concepts, and our perception of them rather limited, can we be sure that our current limited definition of time and space is accurate enough in relation to the totality of human consciousness, and does this mean that the idea of time travel is absurd or feasible?

Now, I am not talking about time travel in the science fiction sense, I am not talking about building time machines. I am talking about time travel in the mental or psychic sense. The idea that we can transcend both space and time with telepathic mental powers using an unknown sixth sense, and project our 'awareness' to distant locations in space and time from the comfort of our physical bodies, has been a concept integral to mystical thought and opinion for thousands of years.

Can human beings perceive beyond the space-time-continuum as we know it, and see or sense events occurring at distant points in either space or time?

Before humans developed their intellectual skills, reading, writing, communication, language and so on, and when we were at an early stage in our evolution, we still retained the psychic connection to nature that we can still observe in animals and plants. We have already seen that plants can telepathically receive thoughts from humans, and in his book 'Dogs That Know When Their Owners are Coming Home and Other Unexplained Powers of Animals', author Rupert Sheldrake documents hundreds of verifiable cases of animal telepathy where ordinary household pets have shown a psychic sensitivity to their owners, which transcends normal explanation.

From dogs who sniff out undetected cancers, or warn their owners of impending epileptic fits, to the thousands or ordinary household pets who somehow seem able to sense unseen dangers yet to come, all show some form of awareness of impending events and possible dangers unknowable by any normal sensual means.

From predicting earthquakes, to sensing and predicting illness - and even death in their owners, or sensing danger to members of the pet's family - even though they may be thousands of miles away - animals, including cats, dogs, horses, snakes, rabbits and chickens, have proved that they are capable of space-time transcendence, and quite capable of provable telepathic and predictive feats, such as predicting Earthquakes.

On 26th September 1997 a violent earthquake struck the town of Assisi in Italy. However it was later confirmed that the night before the earthquake nearly all the animals in the town had started acting strangely. Dogs were howling, others became listless or restless. Cats went into hiding. The flight of birds became erratic. Rats fled the sewers. In China, it has been noted that before seismic activity, snakes will come out of hibernation, and farmyard animals such as pigs and chickens start behaving very strangely. Animals are able to somehow sense these earthquakes before we do, despite our immense technology!

In our ancient cultures animal behaviour was closely observed and then interpreted for signs or omens. These omens remain integrated into 'old wives tales' and nature folk law and some still survive to this day. To the Shaman, however, animals were powerful spirits to be respected - the keen sight of the eagle, the power and grace of the big cat, the wisdom of the snake, the brute force of the bear, the cunning of the fox, the loyalty of dog, all animals were believed to have special powers and to be able to teach and guide man.

The bonds between humans and animals survive to this day, and in households where there are pets, there is often less sickness. Animals give us unconditional love. Dolphins too, can sense sickness and disease and in the Florida Keys there exists one of many programmes where animals are helping to heal handicapped children and adults.

Dr. Sheldrake believes that there are extremely strong emotional bonds between people and animals, and can be likened to what he described loosely as being like a long rubber band which connects them, each to the other, and by which either will be affected by vibrations or events at the other end of the cord!

This strong auric or morphic connection would explain telepathic awareness of those with whom we have strong bonds, one or the other sensing the 'vibrations' at the other end of the cord, but how does it explain telepathic awareness of those with whom we have no particular feeling, no particular bonds?

From what we have already discovered, it would appear that we are bonded to each other and the whole of the universe at the quantum level in some invisible and mysterious way. Consciously, and emotionally some bonds are obviously stronger than others, and some messages will get through easier than others, depending on the wavelengths on which the sender and receiver are operating.The thoughts, the mind and the brain all work on different wavelengths, much like a radio or television set does. Somewhere in our minds, we are both sender and receiver of thought, sound and light waves, and telepathic to each other to one degree or another.

In the same way that you cannot see or hear the television or radio waves passing through your space right now, you know that when you turn on your radio or T.V. the image will be made visible. It is the same with the mind and the sixth sense. The thought waves we emit are invisible. We merely need to be able to 'tune in' to the frequency of the individual to become aware of the thought waves that are constantly being received on the auric field.

If humble animals can sense or forewarn of distant or future events and transcend the apparent physical limitations of time and space with their invisible connections to future events, surely, it would be a travesty to suggest that us supposedly 'superior' humans do not even possess the powers and gifts that these other 'lesser' creatures do? However, in the seventy or more cases Dr. Sheldrake investigated of animal to human telepathy, he only found five cases of human to animal telepathy. All of them were women.

Is the sixth sense a sense that all biological species share? Have we humans merely lost or suppressed the gifts of extra sensory perception, sacrificed at the altar of evolutionary history in pursuit of material, scientific and intellectual accomplishment?

Do humans still possess this gift, and if so, how does it work?

Telepathy & ESP

'The inward man is not at all in time or place, but is purely and simply in eternity.'

Meister Eckhart.

Have you ever walked towards the phone before it started ringing, or knew instinctively who was calling at the other end? Perhaps, even more dramatically, you have sensed that something was happening to a friend or loved one, even though they were at a great distance, or had a premonition of some future event, foretold an event yet to come?

Whilst some might argue these are merely co-incidences, when we study quantum entanglement, we can see that everything is connected everywhere and anywhere in the universe regardless of time or space.

You can visualise this to make it clearer. Firstly, see a beautiful, intricate spider's web, see how still it is? But when a hapless victim lands on the web, what happens? The whole web vibrates instantly. If, indeed, the 'quantum web' that links all things is similar, any disturbance on the 'quantum web' will vibrate across the whole thing simultaneously.

The agitating factor is thought. One might think it is deeds that cause the agitation, but deeds are always preceded by a thought. Thoughts vibrate across the quantum web simultaneously. They do not need time to travel through. This is how telepathy is possible. A part of us is connected to the quantum web, the energy field, through the identical energy that pervades our own being.

Simply put, we are all connected and the part of ourselves which perceives those connections exists at a subatomic level - at a quantum level - and is able to translate them to events or experiences via the subconscious which is then passed to the conscious mind and experienced as 'intuition.' Break the word up, 'in-tuition'. Tuition from within. Knowledge from within.

Indeed, groundbreaking new research now suggests that all living cells could have the molecular machinery for a sixth sense! In an article by Carly Cassella in 'Nature' magazine, the author suggests new data which indicates that every animal on Earth may possess the molecular machinery to sense magnetic fields. The new finding suggests that 'magnetoreception' could be much more common in the animal kingdom - which includes us - than we ever knew. If researchers are right it may be an astonishingly ancient trait shared by virtually all living things. This doesn't mean we can consciously perceive them, because it is an invisible force, but it does suggest that all living cells might, including ours.

Before the idea of quantum entanglement became mainstream, scientists were already considering this idea. In 1964, physicist J.S.Bell presented mathematical proof that confirmed the theory that subatomic particles are indeed connected in some way that transcends space and time, so that anything that happens to one particle affects all other particles simultaneously. These connections are immediate and do not need time to travel through. I believe these particles are known as Tachyons.

Although Einstein thought it impossible for any particle to travel faster than the speed of light, in Bell's theorem effects can be seen to be 'superluminal', in other words, faster than the speed of light. The idea that particles can travel faster than light - although in direct contradiction to Einstein's theory of relativity - certainly appears to be substantiated in recent scientific experiments.

In September 2011, researchers working at the Gran Sasso facility in Italy published proof that muon neutrinos travel faster than light - sixty billionths of a second faster in fact. However much faster this may be in universal terms, however, this is still not as fast as thought - which appears to be instantaneous regardless of either time or space.

Perhaps the superluminal inter-connectedness of all matter that is apparent at the sub-atomic level that Paul Davis refers to as the 'Superforce', is the means by which this instant 'telepathic' communication occurs. If we accept the omnipresence of some form of 'supreme universal consciousness' which appears to be some indescribable form of energy - perhaps pure mind itself - which is, in effect, and at the deepest and most profound levels, the totality of all things, all things being but One - then we may have to drastically alter our current world view of consciousness. In effect, we are no longer individual beings in a vast and mysterious cosmos, but part of the wholeness, and the oneness of creation.

Now, you will remember the transcendental experience of 'oneness with all things' that was frequently reported during altered-states. If the whole of creation is but one giant organism then perhaps there is no separateness, only the illusion of separateness, and in essence we should be able to inter-connect with any other part of the creation - ourselves - anywhere, any time and any place. This is, in essence, because we are already there! Essentially, we are everywhere all at once, certainly at the subatomic and subconscious level.

New theories in quantum entanglement suggest that particles link together in a certain way, no matter how far apart they are in space. As we have already seen, a subatomic world exists within each of us which is connected to the universe via the electromagnetic force field (auric body) and as we have already observed, anything that happens anywhere in the universe, regardless of time and space, is felt instantly on the electromagnetic (auric) body. This is received as waves or vibrations.

The transference of this information from the electromagnetic field (quantum self) to the unconscious mind, and then to the conscious mind, is, in universal terms, a mere formality. The unconscious, having received the data from the subatomic structure of its own electromagnetic force field, merely has to pass it on to the conscious mind, which is when you and I become aware of it. This is usually done in symbols, and these symbols may be audio or visual, externally, or internally perceived.

One psychic I interviewed maintained that she can see 'what others can't see'. By shutting out all outer impressions, by closing the eyes and quietly watching images or symbols that she says are similar to dreams, she can pick up information not available in any other way. She says that she also simultaneously feels sensations in her own body which reflect what is happening to someone else, peace, fear, anger, joy and so on and by giving herself to 'a force' and trusting in it, she becomes a channel for that force and can interpret it accordingly. She said that she trusted the information would be relevant to the person at that moment.

She also further suggested that she was able to merge her own energy field with that of the subject, and that she could 'become one' with it, and in that way access inner knowledge and information. 'It's even more complicated than that' she adds, 'Because what's going on for them, is part of me if you like, and the energy field at that time is reflecting Past, Present and Future. Distance is no object. Distance is physical and it is simply not relevant regarding the energy field.'

Before we, collectively as a species, developed complex

analytical and language skills (left brain) as a means of communication with each other and our environment, we would have relied on our (right brain) skills at assessing environmental factors by telepathic means. Because other species of animals did not take the evolutionary path to the development of left brain activities, they are still very much in tune with their right brain, and highly telepathic.

Do humans still possess telepathic powers? Is there any scientific evidence to support the existence of a sixth sense? Yes, there is.

At the University of Edinburgh in Scotland, remarkable results involving scientific research into telepathy and ESP are currently being made. In tests involving randomly selected volunteers in fully controlled experiments a 45% success rate is being maintained, defying all logic!

In scientifically controlled experiments where the 'sender' and 'receiver' are separated in strictly controlled conditions, strong evidence of the existence of genuine telepathic powers appears to exist. In strictly controlled scientific experiments, the 'sender' is asked to select one of a series of images and 'beam' this image to the 'receiver'. Overall studies showed an above average score in experiments and correct identification of selected imagery in the so called 'Ganzfield' experiments have success rates far above what would be expected purely by chance.

When I asked the late Professor Robert L. Morris, who then held the Koestler Chair in Parapsychology at Edinburgh University, about his own personal views regarding his experimental data, he replied 'I have no strong views on telepathy in humans. We do, however, have increasingly sound evidence that humans can interact with their environments through some additional means of communication. But the results are still very 'noisy' and this makes it harder to study these effects systematically. We may also be dealing with more than just one new ability. In our own research we have obtained results both with, and without the existence of an agent, or sender. Thus our evidence may be evidence for some sort of general ESP and not necessarily for telepathy in particular. I have a personal interest in learning as much as possible about human communication and its full range, but I try to avoid terms like 'proof' as technically, proofs are only obtained by logicians and mathematicians. In science there is always room for at least some doubt.'

An intense and unusual study of the sixth sense and the use of telepathic powers was also conducted for over twenty years by a most unexpected source, the United States government - with extremely positive results! Under the direction of the CIA (Central Intelligence Agency) and the DIA, (Defence Intelligence Agency) at Fort Mead in Maryland, the main objective was to use psychics to gather sensitive information, and penetrate intelligence targets with their sixth sense – with mind power alone. This was known as the 'Remote Viewing' programme.

However, the remote viewing psychics had not been randomly selected, as in most studies but had been specifically selected because of their already known history of psychic experiences. Some of these included out-of-body and near-death-experiences, and in a surprisingly high number of instances, claims of alien abductions and contact with UFOs! One remote viewer began to have psychic experiences after a head injury that he had received in combat.

Ingo Swann, a top psychic who worked on the project for many years believed that anyone could develop these skills with a minimum of training and Dr. Keith Harary - another remote viewing psychic - said that he got his impressions through 'images and feelings, almost subliminal, as though one part of my mind was trying to communicate with the other' - even feeling, smelling and tasting perceptual parts of the experience.

Another psychic involved on the Remote Viewing

project said that it was as if he had a scanner, or as if he opened a 'channel in his mind' to receive this data. Another senior member of military personnel involved in the Remote Viewing Programme said; 'We know there's something there, we can't explain it, but we know it works.'

Remarkably, the most recent scientific research published in 2011 has now proved that humans really do have a sixth sense that lets them detect magnetic fields. Research has shown that humans may have the same innate sense of Earth's magnetic field that has long been known to exist in animals. This suggests that the 'sixth sense' does exist in humans but we might not be aware of it. Neurobiologist Steven Reppert of the University of Massachusetts Medical School quoted: 'It poses the question: 'Maybe we should rethink about this sixth sense?''

An earlier study from Oxford University also found that birds may be able to 'see' the Earth's electromagnetic field as they fly through the sky. Tests showed that different reactions are produced in the eyes of all birds depending on which way the Earth's electromagnetic field spins. The new study was published in the journal Nature Communications.

However, it has long been asserted by psychics that the more sensitive rods and cones in the corner of the human eye are what enable humans to 'see' beyond the physical world. Those who claim to see auras, or even ghosts, which are more ethereal than physical, may be using this part of the eye.

How do we as individuals access our (6th sense) and our telepathic powers?

To open the mind, you must firstly shut the mind! Although this may sound like a strange Zen conundrum, it is really quite simple. Firstly, you must shut down the part of the mind that perceives the material world (left brain), then, using simple techniques to induce alpha waves, you must open your mind, the part of your mind that is connected to the world of the invisible (right brain).

This information is being continually received at the subatomic level on the electromagnetic force field. All you have to do is still the conscious mind (left brain), focus on the subject or object in question, and listen to the thought impressions coming in through the (right side) of the brain (unconscious). You have to become receptive to it. This sounds simple, but it takes time to learn to focus and concentrate so profoundly on one thing at a time.

Where and what is the organ through which this mental reception occurs? How and where do we receive these psychic impressions? Exactly where is this 'scanner' that is frequently referred to?

In the centre of the forehead is a gland called the Pineal Gland. This gland is also known in Eastern and Esoteric literature as the 'Third Eye Centre' - sometimes known simply as - the 'Mind's Eye'. We have all heard the expression 'I saw it in my mind's eye'. Could the 'mind's eye' be the 'third eye', the sixth sense? In our aquatic cousins the Dolphins, this glandular centre in the middle of the forehead contains a small magnetic field. It is from this point that they send and receive information in their oceanic homes.

Is the proverbial Third Eye Centre in essence a magnetic sensor that passes invisible universal, electromagnetic and subatomic information to and from the individual life form and connects us to our inner and outer spiritual being?

Christ himself refers to the third eye centre when he says; 'Let thine sight be single, and thine whole body shall be full of light'. This is a direct and distinct reference to the single eye - the third eye centre - and is as profound as it is proverbial. However, as Christ appears to have travelled extensively in India, he would have been familiar with Eastern concepts and would have learnt about the third eye from the Yogis and Brahmins.

So, is the third eye centre (the pineal gland) the mind's eye which connects us to the inner world of light, and is it through our aura's (electromagnetic force fields) that we receive these subatomic impulses? By shutting the physical eyes we automatically begin to induce alpha waves, by shutting the eyes and focusing on the third eye centre, or pineal gland, Alpha waves are induced even faster. This is why people who meditate focus on the third eye centre.

As we have already seen, we are all capable of altered states, and it is in these moments of altered consciousness that we can access events in time/space that are beyond our immediate sensory perception. Despite the fact that millions of case histories and evidence worldwide suggests humans are capable of some form of telepathy or ESP., the scientific community in general continues to dismiss the existence of a sixth sense with characteristic zeal.

The evolutionary trend on Earth however, is distinctly in the direction of the development of our mental, telepathic and psychic powers and thousands of people every day all over the globe are reporting psychic experiences as yet not fully explained by science.

However, with or without scientific approval, we are all still unconsciously in tune with the whole Universe, and we have only to develop these skills once more, to benefit from the insight and wisdom that we might acquire by listening to our inner voices.

Prediction and Prophecy.

'In most mystical magical and religious traditions, we find not only the key concept of divine inspiration that cuts through all the trappings of religion and causes direct utterance of truth, but also some very specific techniques.'
R.J.Stewart 'Prophecy'.

Whilst Telepathy, E.S.P. may enable us to instantaneously pick up information from distant sources, despite the obvious lack of any normal means of external or sensory communication, an even more interesting form of psychic or telepathic communication exists which, it would appear, not only allows us to see distant locations in space, but also enables us under certain circumstances, to see distant locations in time.

In 'The Bible Code' author Michael Drosnin documents the story of an Israeli mathematician Dr Eliyahu Rips, one of the world's leading experts in group theory - a field of mathematics that underlies quantum physics - who has discovered a hidden code in the Bible which appears to predict details of events which were to take place thousands of years after the Bible was written. These include World War II, the Moon landing and the recent collision of the Hale-Bopp comet on Jupiter! The Bible code also accurately predicts the assassinations of President Kennedy and Yitzhak Rabin.

How is this possible, how could our recent history have been known to those living over two thousand years ago? Does this imply that the future is not random, but that the future is already written?

An ancient form of Hindu astrology originally practised in Tamil Nadu in India, is based on the belief that the past, present and future lives of all humans were foreseen by Hindu sages in ancient times and written down as Palm Leaf Manuscripts. These manuscripts even predicted the exact date and time that each person would call to have his 'destiny' read, or deciphered.

Although the idea of a pre-written destiny is abhorrent to Western philosophy 'what about my own free will?' being the first and most indignant objection - the theory of predestination cannot be discounted entirely and forever in view of the mounting evidence which suggests that the future is entirely predictable.

Now, earlier we saw that the two identical leaves that were subjected to Kirlian photography had their own futures already written on their little energy, or auric fields. It has, in this way, already been scientifically proved and verified that events firstly happen on the invisible energy field before they manifest onto the physical plane. Could it be, therefore, that future events already exist on the energy field, merely waiting to be materialised on the material plane and that somehow, we are sometimes enabled to decipher the code?

One cannot think about prediction and prophecy for long without thinking of that enigmatic mystic and seer Nostradamus, who, in the l6th Century, over five hundred years ago, by a combination of magical rituals and astrology, was enabled to predict distant future events with stunning accuracy. Amongst his many prophecies, he accurately foresaw the Great Fire of London, World War II, and the assassination of President Kennedy.

If the future were not already written and pre-ordained Nostradamus would only have seen a million future alternative realities and been unable to see the future as singularly as he did. And how is it that both the Bible Code and Nostradamus foresaw the same things - World War 2 and the assassination of President Kennedy amongst them?

If, as we have already seen, material creation is a manifestation of an invisible energy field, on which changes occur before manifesting onto the physical plane, the energy field itself would already be generating, or have generated 'the future' before 'the future' was actually visible, before it was materialised.

Some people can tap into it and become aware of it intuitively.

When he was just 12 years old, Sir Winston Churchill, who led Britain to victory in WW2, was asked what he wanted to be when he grew up. Churchill replied 'I don't know. But I know that one day this country will be in great peril, and it will fall to me to save her.' How could he know what was to come 4 decades into the future?

Is there a written destiny for us all?

If we accept that we as individuals contain this energy field, and if we also accept that our planet has an invisible energy field which exhibits changes before manifesting onto the earth-plane, then we are only one step further from considering that the universe itself also might have an aura, or electromagnetic force field, (the anti-material universe?) and that events that will occur on the physical dimension have already been created on the aura of the entire universe.

Furthermore, if we go back to astrology and remember that by mathematically predicting the inter-relationships of planetary energy, that we can predict the manifestations of that energy in relation to living forms, then we need to consider the fact that the future is already written.

Despite the fact that many religious texts are full of such predictions and prophecies - and this includes the Bible - religious leaders are loath to accept these phenomena as real for the 'ordinary' people and regard them as the exclusive domain of ancient prophets and holy men. Yet today's mounting evidence tells us otherwise. It tells us that we are all able to access higher forms of consciousness and experience the transcendental and mystical, given the right environment or training.

Of course, we are not all prophets and soothsayers, and although many people may experience spontaneous revelations which enable them to foretell of future events, some people deliberately seek guidance for the future consulting astrologers, tarot readers, psychic readers, clairvoyants, rune readers and so on, in an attempt to gain understanding and guidance for the future. This is nothing new as people have been seeking wisdom from diviners of one sort or another since humanity began to communicate.

We have always sought help and guidance in this vast and mysterious world that we call home. This is entirely understandable. That we should seek forewarning of impending events is no sin - forewarned is forearmed - especially in terms of earthquakes and other natural disasters. The universal need to know something of what the future holds and how we can best prepare for it, survives to this day.

Prophecy and prediction - long denounced by western religions as heretic and blasphemous - obviously discount and conveniently ignore the fact that Christ himself was himself adept at prophecy. Indeed, he tells Peter shortly before his betrayal 'Before the Cock crows you will denounce me three times'. According to the New Testament, this is exactly what happened. Although the Church and 'Christian teaching' has maintained a deliberately hostile attitude to mystics and the mystic arts throughout most of their history, they seem to forget or ignore the fact that Christ himself was one of the greatest mystics of all time.

The nature of space and time is still subject to scrutiny and definition however, and although time and space to the conscious mind is a fixed reality, time and space, so far as the unconscious is concerned, do not exist at all.

Dreams and Premonitions.

'By paying careful attention to the unconscious, as manifested in dreams and fantasy, the individual comes to change his attitude from one in which ego and will are paramount to one in which he acknowledges that he is guided by an integrating factor which is not of his own making'.

Jung. 'The Essential Jung'.

The purpose and meaning of dreams has long since fascinated humanity and there are many throughout history who have shown abilities to perceive future events, through dreams, out of the immediate time/space scenario.

Amongst ancient primal peoples, dreams were regarded as having great meaning and significance. So much so, that he Native American Indians regularly held 'dream festivals', where the dreams of the tribe members would be pooled, and the contents scrutinised as a means of detecting omens and portents for guiding tribal affairs.

Although modern society holds little regard for dreams except perhaps in the context of personal psychological analysis, many great civilisations and cultures, including the African, Aboriginal, Indian, Chinese, Mesapotamian, Egyptian, Arab, Greek, Roman and European, have sought guidance in personal affairs, and affairs of state, through the meaning and interpretation of dreams.

In dreams, as you will know for yourself, Time and Space simply do not exist. In dreams we can be anywhere we wish to be. In dreams we experience instant reality, the instant manifestation of our thoughts, our will and our desire.

Although fashions and trends in Western thought have changed considerably in relation to their materialistic development, the Shaman retain their connection to the natural world and to the dream world.

The Aborigines of Australia whose culture dates back over 60,000 years, still follow their ancient traditions to this day and are a vital connection to a rapidly dwindling system of primitive and arcane knowledge. The Aborigines believe that 'The Dreamtime' is a universal energy where past present and future exist simultaneously and which can be accessed through dreaming.

Amongst the North American Indians, who, like many early tribal cultures, shared similar beliefs to the Aboriginals, Crazy Horse, an Oglala Sioux, and a friend of Sitting Bull, believed that the world men lived in was only a 'shadow' of the real world. To get into the 'real word' he said, he had to dream, and when he was in the 'real world' matter seemed to dance or vibrate. That Crazy Horse should perceive matter dancing and vibrating would suggest that the world he was experiencing in the 'dream world' was the pre-physical energy world, in which all material forms are originally created.

Could it be that the dream world is the 'real world' and that what we consider the 'real world' is just a dream? In the words of the Chinese philosopher Lao Tse, 'Am I a man dreaming I am a butterfly, or am I a butterfly dreaming I am a man?' The Butterfly has long been an ancient symbol for the soul.

Although the modern western scientific view of sleep and dreams is equated with the brain needing to shut down at night to 'download' data acquired during the day by the conscious mind, a far more profound theory of why we need to sleep has existed for thousands of years amongst mystics.

During sleep, it is said, the soul or spirit - the essence of consciousness of the individual (electromagnetic force field, aura, astral body, soul body) - leaves the body, to go to the spirit world, (pre-physical world) and the dreams are the records of its travels in another dimension.

Can the soul leave the body at night? At the Maimonides Medical Centre in New York, rigidly scrutinised scientific trials have been conducted in which sleeping volunteers were asked to collect information from a package or envelope which was hidden out of his or her sight on a shelf out of their physical reach. The volunteers were then asked to find the envelopes in their dreams whilst asleep and to report the contents on awakening. An overwhelming number of people were successfully able to describe the contents the following morning.

The following is an interesting case history that would appear to substantiate the concept of dream travelling where a distant unknown location is described in a dream: 'I clearly remember 'flying' through the house, noting the colour of the decor as I went. The hallway was rose pink, and the living room, which was separated from the hallway by glass doors, was a pastel green. As I exited the front door - flying straight through it and still flying at about seven feet above the ground - I noticed that the whole of the front of the house was painted white, which is quite unusual where I lived, and that there were two identical, extremely tall fir trees on each side of the entrance. My friend later confirmed that the description I had given was exactly that of his house.'

Although this dream had no significant meaning in terms of future events, the following is another casehistory worthy of mention. 'For three nights running, I had the same strange and disturbing dream. In the dream, I seemed to be flying over a large city - at least my perspective of the scene was definitely from above, rather than ground level. The city itself had many tall buildings, but unfortunately everything looked wrong. The city was partly in flames, crumbled buildings and rubble were everywhere, and the cars seemed to be all over the place, motionless, and some, severely crushed or dented. There was a definite atmosphere of doom and disaster. These dreams were exceedingly obvious to me, because they did not follow my normal dreaming pattern, and had greatly disturbed me. So much so that I mentioned this to several people. On the fourth day, however, I awoke to the following newspaper headlines 'Giant Quake Hits L.A.'. Now I suddenly understood what it had all meant.'

This case history of dream prediction and premonition is not unusual. Many people report having premonitions before dramatic world events, and the attacks in New York on 11th September 2001 were no exception.

David Mandell, otherwise known as the 'Seer of Sidbury Hill' is a former engineer from London and a psychic artist who claims, like Chris Robinson, to be able to predict future events through his dreams. One night, five years before the drama unfolded in New York, he had a vivid dream. In the dream he witnessed two skyscrapers being hit by planes and crashing to the ground. He painted it and took it to be Bank to be photographed by the bank clock, which displayed the day, year and time, so that his predictions could be verified at a later date. The date on the bank clock was 9.11.96. Exactly five years later, to the exact day, the world witnessed the tragedy unfold at the World Trade Centre. Many people claim to have dreamt about this, and other forthcoming events either weeks, months, or even years before the events occur. This challenges our very notions of time and space.

After precisely predicting several major dramatic events in the U.K. Chris Robinson has become well known for his persistent and accurate dream predictions to the extent that he now regularly helps the Police in their enquiries. Interestingly, Chris said his powers developed suddenly after his dead grandmother appeared in a dream and told him that burglars were trying to break into this car as he slept.

The next day a neighbour told him that he had indeed seen a group of youths trying to break into his car, and that he had chased them away sometime in the middle of the night! How had Chris's grandmother, who had been dead for years, see what was happening and how could she communicate through dreams?

Are dreams then not dreams but the record of a reality that is happening on an altogether different dimension to that which we perceive with our conscious minds?

In his book 'An Experiment with Time' author W.G. Dunne recounts his many experiments with dreams, having discovered as a youth, that he was able to dream of future events - although not necessarily of great or dramatic events. In doing so, he realised that in the dream state since time and space did not exist as we know it, that one was equally able to travel both backwards and forwards in time and to perceive future events. He writes 'That the universe was, after all, really stretched out in time and that the lopsided view we had of it, a view with the future part unaccountably missing, cut off from the growing past parts by a travelling present moment, was due to a purely mentally imposed barrier which existed only when we were awake, so that in reality the associational network stretched, not merely this way and that way in space, but also backwards and forward in time'.

Could it be that at night, our soul or consciousness leaves our physical bodies and is then free to travel to some higher dimension of time/space at will?

Have you ever had a flying dream, or woken up in the middle of the night with a jolt? Or have you perhaps woken up in the middle of the night totally unable to move? This is known as sleep paralysis. Although there is no conventional medical explanation, spiritualists believe that this occurs because you - your astral body - is still outside the physical body, but close to it just before re-entry. Anyone who has had the latter of these experiences knows that only a supreme effort of mental will can get the soul back in the body. If you have had any of these experiences, the chances are you are already becoming conscious of your own dream travels.

One thing we do know from our own dreams however, is that in dreams there is no time, only the 'Now'. In dreams, there is no space, only 'everywhere'.

In dreams we can fly.

Chapter Five.
EVERYTHING IS MIND

The Out-Of-Body-Experience

'It looked altogether easy, and they tried it first from the floor and then from the bed, but they always went down instead of up. 'I say, how did you do that?' asked John, rubbing his knee. He was quite a practical boy. 'You just think lovely wonderful thoughts and they lift you up. Of course, they have to sprinkle you with fairy dust first!'

'Peter Pan'. J.M. Barrie

People have always been fascinated by the thought of human flight. There is, perhaps, something in the human soul - perhaps some distant memory, perhaps some unconscious understanding that somehow we can transcend space-time and gravity and literally fly. Of course, we would need an aeroplane to fly with our physical bodies, but there are many myths and legends in every culture that talk about flying. Not with the body, but with the spirit, with the mind.

 For most of us, the ability to fly is usually restricted to our dreams, but there are many cases of people who report being able to fly - out of their bodies - when they are not asleep but awake. How is this possible?

 The concept of a soul which can and does leave the body during sleep and finally at death is a concept shared by many religions and in their book 'Future Science', authors Krippner and White list over 69 primal religions which have recognised the existence of this soul body.

 The soul is considered divine and non physical and part of infinite consciousness which both precedes and creates the material form itself. The consciousness of the individual is linked to the higher energy body which exists within the physical body, and which we have so far only loosely referred to as the 'electromagnetic force field', 'aura', 'etheric body', 'astral body', 'soul body' 'quantum body' and so on. Created, as all things are, by supreme energy, or consciousness, each individual soul - human or otherwise - contains this supreme energy that both precedes it, and manifests it. We all exist ultimately at the atomic and subatomic level, and our physical forms, as we have already quite clearly determined, are direct manifestations of this energy.

On October 26th 1959, two American atomic scientists were awarded the Nobel Physics Prize for the discovery of the antiproton, proving that matter exists in two forms, as particles and antiparticles. According to the basic assumptions of the new theory, there may exist another world, or an anti-world, built entirely of anti-matter. This anti-material world consists of atomic and subatomic particles, spinning in reverse orbits to those of the material world.

How can the anti-material particle be explained?

The Bhagavad Gita, one of the oldest scriptures in the world, gives the following description of the non-material particle, of which we are all made 'The non material particle, which is the living entity, influences the material particle to work. This living entity is always indestructible. As long as the non-material particle is within the lump of material energy, then the entity is manifest as a living unit. In the continuous clashing between the two particles, the non-material particle is never annihilated. No one can destroy the anti-material particle at any time -past present or future.' (Bhagavad Gita 2.19)

In Bhagavad Gita 2.13 it states: 'The anti-material particle is within the material body. This material body is progressively changing from childhood to boyhood, from boyhood to old age, after which the anti-material particle leaves the old unworkable body and takes up another material body.'

In the words of St. Paul, in the first Epistle to the Corinthians, 'There is a natural (physical) body, and there is a spiritual body'. We are in essence, two beings, both divine and physical. We have a physical body, and we have a spirit, or soul (astral) body.

Now, do you remember earlier, we talked about the passage from Genesis that tells us that 'God made man in his own image'? The universe itself manifests originally from pure Light, and now, we discover our own true image, the image of our own souls are also pure light. God made all things in his own image! Spirit, manifest into form! And each of us is a microcosmic universe. Starting with light, we condense into matter through the same processes that brought the whole of creation into being!

In essence, we are pure spirit (energy, light, consciousness) manifesting as form, and, as the spirit itself must ultimately leave the physical body at death, there is no reason to suppose that it is not either necessary or able for it to do so at other times! Furthermore, the individual soul that can, does, and will leave the body at death, contains an absolute record of everything that has ever happened to us, everything we have ever thought, said and done.

The spirit body (soul) originally enters the developing embryo ready to be incarnated, and finally leaves the physical body at death when that incarnation is completed. However, there can or may be times during our conscious waking hours when the 'soul' body may suddenly, and without warning eject itself out of the physical body.

This may occur for any one of a number of reasons the most common natural causes being fatigue, fear, accident, shock and trauma. However, the out-of-body experience may also occur during altered states induced either by the ingestion of mind expanding drugs, during meditation, anaesthesia or trance, as we shall see from the following case histories:

Charles Gill, a retired engineer for British Telecom,

was admitted into hospital for a routine operation involving general anaesthetic. The drug was administered intravenously and he was required to 'count to ten'. He suddenly found himself very relaxed and floating close to the ceiling. 'Either I went up to the ceiling, or the ceiling came down to me' he said. 'It was as though my consciousness came out of my body at the navel. It was an irresistible but wonderful feeling of release and freedom. However, the feeling lasted only for a few seconds, then the anaesthetic took hold and the next thing I knew, I was waking up in the recovery room with a wonderful sense of well-being. I was very aware that this sensation was identical to one I had experienced as a young child of about five years, when I would regularly 'fly' around my bedroom ceiling, almost at will, and observe both myself and my younger brother, apparently asleep in the cots below.'

The out-of-body experience during illness is not uncommon and although it could be attributed to the illness or the medication itself, the out-of-body experience can also occur under less dramatic circumstances with no medical condition, drug or anaesthesia administered.

One night, tired and exhausted after an 18 hour working day, a prominent UK physicist (who has chosen to remain nameless due to fear of ridicule from his scientific colleagues) lay on his bed to rest when suddenly he found himself floating out of his body. Although he was completely taken by surprise, his curiosity was aroused, and he experimented for a short time floating this way and that, before he found himself moving out of the window - or rather through the window. At this point, however, a sudden realisation of the magnitude of what was happening to him made him somewhat alarmed and he suddenly re-entered his body with a thud. 'I can't explain it', he told me, 'I'm a Professor of Physics, it goes against everything I've been taught. But I'll tell you one thing, I know it happened!'

The out-of-body experience can also happen

spontaneously in accidents as the following case histories show.

Paul Ford, a mechanical engineer, was driving round a sharp bend, when he suddenly and unexpectedly came face to face with a stationary JCB digger. Too late to stop, he crashed directly into it. In the next instant, Paul said he found himself floating over his body, high above the scene of the accident looking down. When he came to, he found that the driving seat in which he had been sitting had been pushed by the force of the accident into the back seat! He narrowly escaped with his life.

Another case history that also occurred during a car accident involved a 19-year old student. 'When I saw what was about to happen, I immediately passed out. A short time after, it could have been seconds, or minutes, I remember thinking 'that's it, I'm dead, this is what it's like to be dead. Quite honestly, it wasn't a problem, I felt exactly like I normally do, except - I didn't have a physical body! I remember becoming conscious in my unconsciousness if you like, the physical world just did not exist. Not the street, the car, the people, nothing. It was just darkness. I assumed it was the Void. But, in the middle of the void was this light which was moving like a filament in a lightbulb in a sideways figure 8, and in the middle of the light seemed to be me, which I could only, at that moment, describe as pure thought and every time I thought something, the filament of light became activated. It seemed to be pure light. Like I said, I felt just like me, but my body, and the material world were nowhere in sight. At the time of the accident I had never heard of the out-of-body experience, and just took the experience I had for granted. It wasn't until I heard about it, that I realised that was what had happened to me.'

Anxiety can also trigger the out-of-body experience and Melvyn Bragg, now Lord Melvyn Bragg - a highly respected author, television producer and presenter, recalls his own out-of-body experiences, when, as a young boy he was temporarily separated from his parents. His anguish was so great, that he suddenly found himself out of his body flying down the country lanes at treetop level in search of his parents. Upset at the temporary separation, the young Melvyn, unable to physically leave the house, had spontaneously left his body. This happened on a couple of occasions, he reflects, but 'I was frightened that I wouldn't be able to get back into my body. It never happened again.'

Fear of what is happening, and fear of whether or not the projector feels he or she can safely return to his or her body is a common feature in the out-of-body experience.

The astral body' or 'soul body', however, is connected throughout life by an ectoplasmic cord which can stretch literally any distance, and is not finally cut until death. This cord is similar to the morphic or ectoplasmic cords to which Dr. Rupert Sheldrake referred earlier. When we are born into the physical world, the umbilical cord is cut to finalise the separation from the womb. When we die, the cord is cut to finalise our separation from the world.

The out-of-body experience can also occur during altered states of consciousness where there is no fatigue, physical danger, or life threatening scenario as the following case histories show:

Cassandra, now a successful hypnotherapist, was 13 when she had her first out-of-body experience. Staying with friends overnight, she had gone to sleep early, when suddenly she became aware that she was floating on the ceiling, looking down at her body on the bed below. Then, from her vantage point above the room, she saw her friends walk into the room, and then leave again when they got no reaction from her, thinking she was fast asleep. She wasn't. She was watching it all from the ceiling. Then she felt a sudden pull back into her body and went back to sleep. She was not aware at the time what she had actually experienced, only that it was something perhaps a little out of the ordinary. She was to experience this again on several other occasions, being out of her body and exploring her family home from ceiling height on a regular basis going from room to room. On a couple of occasions she experimented with purposefully accessing this out of body state and on others it would just happen spontaneously. One night she was surprised to see - not only her physical self lying on the bed below - but a beautiful silver cord connecting her to her body.

Diane Schufler is an interior designer, married, with two children. During her youth she found herself experimenting with her first joint of cannabis, and after inhaling it several times, suddenly found herself floating out of her body, looking down at herself from the ceiling. Once she had become aware of what had happened, she became frightened whereupon, she suddenly found herself back in her body. She later told me that she believed it was a gift, and that if we could harness the power of imagination, we could really benefit from the experience.

Creative Director David Palmer had his first out-of-body-experience when he tried LSD. To his surprise, and several hours after the initial ingestion, he found himself walking or rather, floating behind his physical body as he walked down the street. Although his first experience was a bit of a surprise, it was the first of many which he later learned how to control. He was then able to project in and out of his body at will. On one occasion he recalls 'I was lying on Hampstead Heath relaxing and staring at the clouds, when suddenly, to my amusement, I found myself peering into the window of a passing plane'.

In the out-of-body experience most people seem to stay quite close to their bodies, but David also believes that has travelled through the 'astral plane' - a higher dimension - in his astral body. This is known as astral projection. David says it can get a bit frightening sometimes when you meet different spirits and entities. At that point however, he often finds himself back in his body.

Todd Routt began having out-of-body experiences as an adult and found that with a few simple techniques, he could control them and 'leave his body' at will. He is so convinced of the life-changing spiritual importance of the experience that he now runs workshops to help other people learn how to astrally project.

He believes anyone can be taught to leave their body an experience he calls the 'Journey of truth'. Todd believes that the out-of-body experience helps us to discover that we are more than our physical bodies and that a greater part of us exists beyond time and space. He also believes that the out-of-body experience can profoundly transform your understanding of yourself and the universe.

However, the experience of fear can also drive the 'soul' out of the body and I have spoken to several women in the course of my research who have reported having out-of-body-experiences whilst being raped. In many cases the experience commenced with a profound emotional desire to be 'outside' of the situation. I will not subject you to their harrowing stories, but suffice it to say that the out-of-body experience is linked to our deepest emotions and survival mechanisms.

I have also heard of many people who have had out-of-body experiences during near-drowning experiences including TV presenter Richard Madeley of ITV's This Morning show who at the age of 17 was washed out to sea by an unexpected storm. He was, of course, subsequently saved.

Many people who practise prolonged periods of meditation also report out-of-body experiences, with the associated feelings of floating or flying and although these experiences may appear to be rather fanciful, unbelievable and totally subjective, two common features suggest that they are not.

Firstly, in the out-of-body experience the subject is often able to accurately describe events occurring in that (or another) location. Secondly, the subject may actually be seen by another person in a separate location from the physical body. The second type of phenomena is known as 'reciprocal'.

One night, Alan Clarke, now a TV producer for a regional UK Television station, left home to drive 50 miles to work on a Radio Show. His flat-mate remained at home. During the middle of the night, however, she had woken up to find him unexpectedly standing in the doorway of the lounge. She recalled 'I was kind of surprised to see him there. He looked, really, just like a ball of light, except that I could see that it was Alan. He started to float towards me, and I said something like 'what are you doing here I thought you were miles away' - but I certainly didn't speak it - it was telepathic communication. It felt quite natural.'

At that point, she fell immediately back to sleep, but awoke the next morning convinced he had come home earlier than expected in the night. She searched the flat, but he was nowhere to be seen, and he did not return until later that day. 'In retrospect' she adds 'I must have been out of my body too, because, from where I was sleeping, which that night was on the settee in the living room, I could not have seen the doorway to the living room, as the back of the settee was blocking my view. I have no doubt as to what happened and it has changed my attitude to life, and more importantly, to death, forever. I have seen the soul for myself and there is no going back. I know now that we are more than physical bodies and that death is not the end.'

This is an unusual case of both out-of-body experience and reciprocal vision of a separate individual who was also out of his body which was, at the time, some fifty miles away! Reciprocal vision of a projecting body is frequently attributed to Yogis who have been seen, on some occasions, in either one, or several locations, entirely separate from their physical body which remained inert in another altogether different location.

If we can leave our bodies during life, do we finally leave them at death?

The Near Death Experience. (NDE)

'For the soul there is never birth nor death. Nor once having been does it ever cease to be. It is unborn, eternal, ever-existing, undying, and primaeval, it is not slain when the body is slain.'

Baghavad Gita 2.20

From the moment we are born there is only one certainty - death. We must all come to this final act. And yet, we know very little about what happens to us when we die. Do we believe our religions, all of which promise some sort of afterlife, or do we believe material science that tells us there is no God, no purpose, only random chemistry followed by personal oblivion?

Whilst religion and science appear to have been historically at odds on this matter, new theories in quantum physics are proving that death is just an illusion. According to a new theory, Dr Robert Lanza, a scientist and theoretician, claims there is no such thing as death (of the self), just death of the body. He argues that our consciousness exists through energy which is contained in our bodies and is released once our physical beings cease, in a process he calls biocentrism.

Dr Lanza says we believe we die because that is what we are taught, but the reality is very different and he reiterates Einstein's famous quote 'Energy cannot be created or destroyed, it can only be changed from one form to another.'

As such, when our bodies die, the energy of our consciousness - which scientists do not fully understand - could continue on a quantum level.

The biggest mistake, in my mind at least, that scientists continue to make, is to believe that consciousness is generated by the brain itself. As we saw in the out of body and near death experience, consciousness exists within the energy, or auric field, of the individual, not the physical form, although obviously the brain itself is the medium between energy and matter, which processes all the necessary activity required to exist in a physical body in the physical world.

Although the idea of a spirit double or soul, is at least 60,000 years old, and has now finally been verified by sophisticated scientific and technological photographic techniques which were first developed by the Kirlians during the 1950's, it is only now, since the remarkable progress in medical resuscitation techniques, that we have had a virtual worldwide deluge of first hand accounts of people who have literally died, and come back to life.

In the USA alone, over one million people admit to having a near death experience in recent years. In nearly 100% of cases a virtually identical experience has been reported, regardless of race, creed, colour, age, sex, or cause of 'death', with several key similarities:

1) The subject, having suffered clinical 'death' - which is fully confirmed by qualified medical staff - reports floating out of body totally free from pain (whatever the cause of 'death') usually floating to a vantage point on or near the ceiling where the subject invariably looks down at his or her own body lying below.

2) The subject next reports flying down a tunnel of Light towards the source of Light, which is accompanied by intense feelings of peace, beauty and love.

3) The subject then appears in a beautiful garden listening to incredibly beautiful music and then either reports seeing or feeling a divine presence, and/or dead relatives who tell them it is not their time.

4) The subject then finds him or herself back in their bodies.

Several other common features of the Near Death

Experience include a sense of overwhelming peace, well being, or absolute unconditional love, a sense of having access to unlimited knowledge, increase in telepathic powers, a life review or recall of important life events, and occasionally a preview of future events.

Although sceptics like Dr Susan Blackmore - who ironically enough has had out-of-body experiences herself - believe that the Near Death Experience is the hallucinations or over-oxygenation of a dying brain this:

1) Has never been clinically, medically, or scientifically verified or proven - or subject to scientific scrutiny, and is only vaguely theoretical, being a gross medical assumption rather than a proven fact.

2) Does not account for the fact that when the N.D.E occurs, the subject, both body and brain, is biologically and clinically dead.

3) Does not account for the fact that the out-of-body-experience, which is identical to the first stage of the near death experience, does not involve the life or death scenario, and may occur under a variety of stimuli and conditions - taking psychoactive drugs, fatigue, anxiety, fever, meditation, accidents, altered states, and so on.

4) Does not account for the fact that when the subjects were literally clinically dead, they were able to accurately describe events that occurred in various other locations during the out-of -body and near-death-experience itself, which were later verifiable by reliable eyewitnesses. (This is known as a veridical N.D.E - in which the subject acquires verifiable information that they could not have obtained by any normal processes).

5) Lastly, the mathematical odds of several million people coming up with an identical story whilst they were clinically dead, is probably several trillion to one!

One noted expert on the near-death-experience is Dr. Peter Fenwick, MB, BChir DPM. F.R.C.Psych, and a Fellow of the Royal College of Psychiatrists. He is also a consultant neuropsychiatrist at the Maudley Hospital, The John Radcliffe Hospital in Oxford, and holds a research post as Senior Lecturer at the Institute of Psychiatry in London. However, It was during his work as a senior consultant at St. Thomas's Hospital that he first heard about the NDE, and he became increasingly curious about the number of reports of near-death experiences he was getting from his patients on a regular basis.

He then began to amass the data and found further evidence that whilst clinically dead, the subjects were able to describe, in accurate detail, events that were happening in a location distant from the physical body itself. He also found amazing similarities in all the personal stories of near-death-experience that he investigated.

Jean Williams will never forget the day she literally died - and came back to life again. Jean, a retired nurse, had just undergone major surgery at the Liverpool Women's Hospital, but at 6.30pm things went, in her own words 'horribly wrong'. The following is her account of what happened.

'On Saturday morning, the nurse came and asked me if she could take my temperature. After the operation, I was bleeding, although no one knew it at the time. She took my temperature and went on to the next bed. I leant over to get my cup of tea and collapsed.

With that, I seemed to leave my body, and I went up to the clock on the wall. I wasn't conscious of walking, I just floated to the clock. It felt real. It felt as though I was really up there and my dead body was on the bed. It wasn't like a dream, it was real life - and by this time a doctor and two night sisters had been summoned - trying to get my blood pressure. A doctor with an Australian voice was trying to get it on the other side. The doctor pummelled me on the chest and nothing happened. A nurse said 'Sorry, I just can't get the blood pressure.' The doctor said 'I'm sorry sister, she's gone - pulmonary embolism.' I just didn't understand what all the panic was about - because I felt free up there by the clock - I didn't understand why they were fighting for me.

Then they pushed my bed with the oxygen cylinder away from the wall. As soon as I saw them taking my dead body away, I thought I better go back, I can't let them go without me, so I seemed to come from my position by the clock back into my body with a thud. Then I had a lot of pain again, but I didn't have any pain when I was out of my body.' (The record of events that Jean Williams recounted during her experience was one hundred percent accurate and later verified by the medical staff in attendance.)

She continued: 'It was 11.30 when I came too, and from 6.30 to 11.30 I was up in what I presume to be heaven. It started with a tunnel of light - a vivid brightness - very serene. I went to this beautiful park with very vivid colours - a chorus of voices singing - trumpets and beautiful music. I saw people I knew, and I saw my father - he seemed to be floating around. In the meantime, my husband was called and told I had died and was asked to come back for the Death Certificate the next day. Then there was a vivid brightness and I saw what I can only presume to be the Good Lord himself. This figure was radiant, absolutely radiant. He was wearing a white gown, he didn't speak to me but he welcomed me, and he was letting the little children play around him. I didn't want to come back from that place - it was so beautiful'.

Derek Skull, MBE, a retired Army Major, married with two children is, in his own words 'a pragmatic, down to earth sort of person'. However, a near death experience that occurred out of the blue when he had a heart attack remains etched in his memory forever. One autumn afternoon, Derek suffered the first of two major heart attacks, and found himself in intensive care at the Kingston Hospital.

On the first day he had an injection of morphine to ease the pain, and he was in his own words 'at peace with the world'. 'Suddenly' he continued 'I took off and floated 'airborne' if you like, into the corner of the room, looking back - there was my body on the bed! 'Good Gracious what is this' I thought, 'but in fact I was looking down at my own body and I had a full view of the ward, and which point I saw my wife in a red trouser suit talking to a nurse, and I thought 'Oh my God - what an inappropriate time to arrive'. The next thing Derek knew was that he was back in his body in his bed, greeting his wife who was wearing - a vivid red trouser suit. 'I know it wasn't a figment of my imagination,' he added 'because it was so clear'.

Some time later Derek Skull had a second heart attack and a second NDE after which he said 'I am absolutely convinced that something - call it your soul, call it what you like, temporarily detaches itself from your body, and goes to a vantage point to review the situation'.

Barbara Lambert was in the Stone Maternity Home after the birth of her first child. She had had a painful drawn out labour and was frightened she would die because the pain was so intense. The next thing she knew, she was two inches away from the ceiling, looking down at herself on the bed. Although she was frightened, after a while it seemed quite natural. 'The next thing I felt as if I was in a long dark tunnel with a very bright light, more brilliant than the sun - absolutely brilliant! I felt it was very natural to be drifting towards the end - it seemed very peaceful and natural. I didn't feel I could go the other way - it just felt like a magnet was drawing me toward the end (of the tunnel).

As I got closer and closer, it was as if I had to stop, I couldn't just go through - and then I saw a small country garden and a man dressed in a tweed suit - no tie- standing behind a fence just a few feet away from me. He recognised me so I felt as if I had some sort of physical appearance.

He asked me why I wanted to die - I can remember thinking 'is it granddad?' I knew he had died, so that I must be on the point of death. He died when I was nine and I felt if I went any further from where I was, I would definitely die. He seemed like he was veering me towards a decision and he said 'who would look after the baby?' and when I said 'I don't know' - that's when I found myself back in my body. It absolutely definitely was not a dream. I couldn't tell you what I dreamt last night, last week, or last year. This was decades ago and it feels like it happened five minutes ago. It's that clear. I have never forgotten it.'

On the other side of the Atlantic in the USA, Raymond Moody Jr. M.D., has been doing his own investigations into the Near-Death-Experience and has collected literally hundreds of stories of NDE experiences. In his book 'Reflections on Life after Life' he relates the story of a middle-aged man who had a cardiac arrest. His story was very typical of other NDEs. 'I had heart failure and clinically died. I remember everything vividly. As things began to fade there was a sound I can't describe: it was like the beat of a snare drum. it was very rapid, a rushing sound, like a stream rushing through a gorge. And then I rose up and was a few feet up looking down on my body. I had no fear - no pain. Just peace. After a second or two, I seemed to turn over and go up. It was dark, you could call it a hole or tunnel and there was this bright light and it just got brighter and brighter. And I seemed to go through it. I was just somewhere else. There was a gold looking light everywhere. Beautiful. There was music. There was a sense of perfect peace and contentment; love. It was like I was part of it.'

In a similar case history one woman describes her own experience, 'There was a vibration of some sort. The vibration was surrounding me, all around my body. It was like the body vibrating, and where the vibration came from, I don't know. But when it vibrated, I became separated. After I floated up, I went through this dark tunnel and came out into brilliant light. There was the most beautiful brilliant light all around. And this was a beautiful place. There were colours, bright colours, not like here on Earth, but just indescribable. Everything was just glowing, wonderful.'

Like all NDE's however, this patient returned from death, but it is interesting to note that in the last two case histories, both subjects were aware of the separation of the physical and astral bodies, and actually heard the sound that is often reported in both OBE's and NDE's, that of an intense vibration which accompanies the separation.

This is not always the case however. In 1985, Australian Chris Parnell suddenly found himself silently standing over his own dead body. Chris – sentenced to 11 years in an Indonesian jail after being wrongly convicted of smuggling hashish – had become the victim of a violent assault by a fellow inmate, receiving five serious stab wounds. He was taken to hospital, where he underwent surgery, but was later declared dead and his body removed to the mortuary.

As he lay on the mortician's slab being prepared for burial he suddenly became aware of standing over his 'dead' body. 'I looked at myself on the slab. I was cold. But the 'me' looking over my body was warm. I kept repeating to myself 'dead is cold' 'alive is warm'. I kept saying it over and over and visualised heat returning into my 'dead' body until I could feel the heat all over me.' Suddenly Chris drew a breath. He had literally come back to life. He eventually recovered from his terrible injuries and although he was promptly returned to prison, was finally released in 1996 after a long campaign by family and friends.

However, children may also experience similar phenomena as the following report shows. In April 2010, three-year old Paul Eicke accidentally fell into the garden pond at his grandparent's house in Germany. Despite a frantic dash to the hospital little Paul was finally pronounced clinically dead. His heart had stopped beating for over three hours. The doctors had just given up when, miraculously, Paul's heart suddenly started beating entirely independently. Professor Lothar Schweigerer, director of the Helios Clinic where Paul was taken, said: 'I have never experienced anything like it. When children have been underwater for a few minutes they mostly don't make it. This is a most extraordinary case.'

The boy later said that whilst unconscious he saw his great grandmother Emmi, who had died many years previously, who had turned him back from a gate and urged him to go back to his parents. Paul said: 'There was a lot of light and I was floating. I came to a gate and I saw Grandma Emmi on the other side. She said to me, 'What are you doing here Paul? You must go back to mummy and daddy. I will wait for you here.' I knew I was in heaven, but grandma said I had to come home. She said that I should go back very quickly. Heaven looked nice. But I am glad I am back with mummy and daddy now.'

Although the medical resuscitation techniques that enabled these people to literally return from death are relatively new in our history, medical resuscitation was known by the ancients using mouth-to-mouth respiration. One such case is mentioned in the Bible in Kings 4:18-37.

Could it be that these early travellers to death and back gave us our first glimpse of Heaven?

However, the most concise scientific theory to date to try and explain the near death experience has come from quantum physics. A near-death experience happens when quantum substances which form the soul leave the nervous system and enter the universe at large, according to a remarkable theory proposed by two eminent scientists. According to this idea, consciousness is a program for a quantum computer in the brain, which can persist in the universe even after death, explaining the perceptions of those who have near-death experiences.

Dr Stuart Hameroff, Professor Emeritus at the Departments of Anesthesiology and Psychology and the Director of the Centre of Consciousness Studies at the University of Arizona, has advanced the quasi-religious theory. It is based on a quantum theory of consciousness he and British physicist Sir Roger Penrose have developed which holds that the essence of our soul is contained inside structures called microtubules within brain cells. They have argued that our experience of consciousness is the result of quantum gravity effects in these microtubules, a theory that they dubbed Orchestrated Objective Reduction.

Dr Hameroff told the Science Channel's Through the Wormhole documentary: 'Let's say the heart stops beating, the blood stops flowing and the microtubules lose their quantum state. The quantum information within the microtubules is not destroyed, it can't be destroyed it just distributes and dissipates to the universe at large.

'If the patient is resuscitated, revived, this quantum information can go back into the microtubules and the patient says - 'I had a near death experience.'

Death and Resurrection.

'Never the spirit was born:
the spirit shall cease to be never;
Never was time it was not.
End and Beginning are dreams:
Birthless and deathless
and changless remaineth
the spirit forever.
Death hath not touched it at all,
Dead though the
house of it seems.'

Sir Edwin Arnold.

Is the near-death-experience a foretaste of death itself? Have our very concepts of heaven and after-life been passed down over thousands of generations and incorporated in early primal religions since the first human had an out-of-body or near-death-experience, and lived to tell the tale?

Although the concept of death in the western world is one of finality, the self being mainly identified with the physical form and brain, the Eastern view is rather different. In Eastern philosophies and religions the overall view is that a human is a mind possessed of a body for expression on the physical plain. The Western view is diametrically opposed. In other words, the Western view is that we are a body, possessed of a mind (defined as the brain) in order to satisfy the needs and desires of the body. According to the latter view, life (consciousness) ends when the body dies. According to the former view, life (consciousness) does not end after the body dies.

In some Eastern philosophies, there is absolutely no difference between the living and the dead. For the Yogi, the Universe is filled with conscious light energy, and the only thing that exists in the entire Universe is pure consciousness.

The soul of the individual is pure consciousness, (or divine energy), and at physical death, the energy itself, which most closely resembles a ball of light, leaves the physical body. The ball of light (soul) contains the personality and memories of every life the soul has had, or maybe even will have, on its journey back to the source.

The concept of a soul that survives physical death is common to all religions, but perhaps the most obvious and potent symbol for the concept of a soul that survives physical death is the Resurrection of Christ. Was the Resurrection of Christ an attempt to communicate the reality of the out of body, near death experience and survival of the soul after death? Was Christ's reappearance after his death, publicly demonstrating to us the truth of the soul's survival after physical death?

Although many traditional Christian scholars believe that the Resurrection of Christ was a physical one, a bodily one, I personally, very much doubt this interpretation. Indeed, as we have already seen, Church authorities are now equally dubious about their own definitions of the scriptures and Cardinal Hume has said 'we cannot be sure the resurrection actually happened.'

Of course, if the Church cannot even be sure that The Resurrection 'actually happened', then they must also surely admit to being equally unsure that the Resurrection itself was absolutely physical, as they have always vehemently maintained and not spiritual, as the near death experience would suggest.

Since the Gospels themselves were not written until many years after Christ's death, to what extent do they fully reflect the facts, and what political motivation might the early Roman Church have for re-interpreting the original scriptures. Power and control over the masses?

Given the current evidence, I believe that the symbolism of the Resurrection was absolutely spiritual. Christ left his body and was seen, in spiritual form in his 'soul body'. Is the Resurrection of Christ unique, or do we all survive physical death?

At the University of Virginia in Charlottesville, USA, Professor Ian Stevenson has been studying a large number of reports dating back forty years of cases where someone who has just died, or is about to die, appears to a close friend or relative- exactly as Christ did to his followers after his own death.

One of the earliest recorded cases of this phenomenon, however, occurred in England in 1250, when the Abbess of Lacock saw the apparition of her son William Longespee, who was at the time thousands of miles away on a Crusade. She told her friends of the incident, but no one believed her. Six months later, a messenger arrived from Egypt to tell her that her son had been killed - at exactly the same time his ghost had appeared to her.

The near-death-experience would certainly seem to suggest that we all survive physical death. As we saw earlier, if consciousness really can exist out of the body, and can move to distant locations at will, there is no reason to suppose this to be any different after death itself when the cord is finally cut.

The concept of a soul that survives physical death is present in all the world's religions bar none. In Buddhism, the attitude to death can best be summarised in the words of the Dalai Lama: 'As a Buddhist, I view death as a normal process, a reality that I accept will occur as long as I remain in this earthly existence. I tend to think of death as being like changing your clothes when they are old and worn out, rather than as some final end.'

The soul can be considered to be a conscious form of light, or energy, which has characteristics of both the highest forms of energy within the matrix of the self, and the characteristics of the individualised forms inherent in the self.

The 'soul' is the spark of God within all things, which is composed of this higher form of energy, and which carries the mark of the individual's present, former and future lives, and stores all the memories of these lives within the higher energy bodies.

We are so used to considering ourselves as purely material form, that we forget that the spiritual form within the physical body is the higher, and immortal, infinite energy that is responsible for the creation of all life. This 'ball of light' has manifested itself as you, to experience the universe from your own unique perspective.

No one but you can see the world the way you see it, or from where you see it. You, and your perception of the Universe are absolutely unique, and this pattern of perspective will be carried within the individual form throughout eternity until it merges with the prime source.

Visions of Paradise.

'And Jesus said to him, 'Assuredly, I say to you, today you will be with me in Paradise'.
The Gospel of St. Luke. Verses 23.43

Although the idea of higher spiritual dimensions has existed within spiritual works for many thousands of years, it is only with the discovery of Superstring Theory, that we find some literal scientific evidence that higher dimensions may be realities and not fantasies after all.

Where are these higher dimensions? Do they exist in some far off place? No. All the dimensions occupy the same space they simply exist at different levels of vibration. They all co-exist within the same space.

In the material world, which is operating on a relatively slow vibration, we perceive only the material forms that are vibrating at the same frequency as ourselves. Go up a vibration, and you are on the next dimension, the astral plane, which exists on a higher atomic rate of vibration, where thought forms, spirits and souls exist, and who float and fly about in their low gravity home.

Their dimension is as real as this one. Like radio stations that can exist on different wave bands and simultaneously broadcast without interfering with each other, each dimension has its own wave band and is, to all intents and purposes, invisible to the others.

Now, according to the overwhelming number of reports and memories of near-death-experiences, which we have become so familiar with, one of the three most common features of the NDE, is that after leaving the body and observing it on the bed (or where-ever) below, the soul then travels down a tunnel, at the end of which is the most brilliant light.

As well as seeing the Light people report hearing the most glorious music and beautiful colours never before heard or seen here on Earth. Beautiful gardens stretching into infinity, and profound feelings of peace, love and compassion never experienced before in human life. Most people who have had near-death experiences also report overwhelming feelings of a divine presence, of Radiant beings of Light and pure Love.

Is it a mere coincidence that the reports of 'heaven' glimpsed in the near-death-experience are identical to religious and artistic renditions? Could it be that our very concepts of Heaven are based on the recall of those who have died, and somehow been brought back from death?

Some sceptics may insist that the visions of paradise experienced by the individual is a 'trick of the imagination' or 'the hallucinations of a dying brain' and that people simply remember pictures and paintings which have interpreted heaven and paradise in this way. But where did the artist get his inspiration from in the first place? Perhaps he or she had glimpsed it themselves in the near-death-experience, or during some plant-induced ecstasy, or in the throes of deep contemplation.

Does Heaven really exist? Do we really survive physical death?

It would certainly appear from the scientific evidence that some form of light does exist within us. We have also seen that this light, our 'soul', can and does leave the body every night during sleep, in the out-of-body experience, and finally at death. This soul does not die. It is eternal.

All great religious texts refer to some form of afterlife. Even before the written word, our ancestors passed these wisdoms through song and legend, down through the generations and the concept of individual soul survival and resurrection is part of the essential teachings of many accepted spiritual and religious philosophies which go back into our pre-history.

However, as far as energy is concerned, it is always creating and destroying life forms in an endless cycle that may well continue for all eternity. The soul is part of that energy and we are part of that cycle. Energy will continue to recreate form. Energy does not stagnate or stay stationary. If Heaven exists - as the near death experience would seem to indicate - is there Hell too?

Jung believed quite simply that Heaven and Hell

existed primarily within the psyche of the self as paired complementary opposites. Certainly 99 out of 100 cases of near-death-experience make no mention of any such dreadful place as Hell. In all cases, the subject experienced going down a tunnel to the most glorious and beautiful, loving and heavenly place imaginable. Many did not want to come back, such were the overwhelming feelings of beauty, love and peace. In only one recorded case was there a difficult experience when the subject could not find the light.

Heavenly music, beautiful gardens, a feeling of incredible love and peace, divine beings made of light - is the mythology of Heaven based on the near-death-experience? Heaven, perhaps, is the source of light itself, in both a scientific and spiritual sense, and it would appear to be that each of us has a direct and personal worm-hole to God, if you like, which connects us to our source, to our God.

This heavenly place, perhaps the highest of all dimensions, is a place of love and light. When we die, we are not judged by a wrathful or vengeful God - we will be judging ourselves. It's called our conscience. Each of us contains spiritual light that is pure spirit. Our souls will judge ourselves. Everything that we have ever done, or thought, every deed and every action will be reviewed once the soul has ascended from the material plane after death, to the next dimension - to the next spiritual level.

The soul will review its own actions and judge itself accordingly. It will then, according to the Law of Karma and Reincarnation, re-enter the earth plane at the appropriate time to continue its learning process on its own soul journey back to Godhead. There is no escape from the all seeing-eye.

So, when we die, do we just go back to Heaven and stay there?

Many believe that we must reincarnate time and time again until we are pure of heart 'Unless ye be innocent as little children - ye shall not enter the Kingdom of Heaven'. (Matthew 18.3) The soul must incarnate time and time again until it is pure enough to return to and reside with God, or exist on the Highest of all dimensions that we call Heaven or Paradise.

Sometimes, however, the spirit, having left the Earth, is not at peace, even in its temporary respite from the round of incarnations, and may wish to return to Earth after its death. These unsettled souls are known as 'ghosts'.

Ghosts, Mediums and Spirit Guides

'Your consciousnesses are separating from your body and entering the Bardo. Appeal to your energy to allow you to see them as you cross the threshold and retain total consciousness. The vivid clarity of the Light without colour and emptiness will appear and envelop you with a quickness greater than lightning. Don't allow fear to make you retreat and lose consciousness. Plunge into that light. Reject all belief in an ego and all attachment to your illusionary personality. Dissolve its Non-being into Being and be free.'

The Bardo Thodal

We are now coming to the second pivotal, most profound, billion dollar question, and it is one that we all have a vested interest in knowing the answer to.

In approximately 120 years time, give or take a year or two, not one single person currently on Earth today, not you, not me, not anyone, will still be here. Neither president nor pauper can cheat death. We will all have passed. We will all have left our earthly bodies, we will have finally shed the mortal coil.

So, what happens to us when we die? Is that the end, or is there something about us that can survive physical death, and if so, exactly what is it?

It is obviously no part of our bodies, not brain, not heart, not liver nor any other part of our physical selves. The bodies we inhabit are just material forms and will have turned to dust. How is it then, that we can - that our consciousness - can survive physical death?

Quite simply, our consciousness is part of the consciousness of the entire cosmos and it is infinite, eternal and indestructible.

We all have this energy inside us, whether dead or alive. We all have a ghost (soul, spirit, quantum self) currently residing inside our bodies. We have a body, but we are not our bodies. Our bodies, like our brains, are material forms and will disintegrate at our death when we leave the body. Therefore the brain cannot be the source of consciousness per se, but rather, the transmitter of it, particularly if you consider the data which appears to prove consciousness can separate from the body in life and finally at 'death'.

Rene Descartes, the French philosopher and mathematician, now referred to as a 'substance dualist', believed that the mind and body were made from different substances, the body was made of a physical substance and the mind of some mysterious non-physical material!

Neuroscientists in general reject this idea, and are still stubbornly of the belief that the brain is the source of our consciousness. This represents a view generally held throughout most of the conventional sciences and neuroscience in particular. We are bodies, with brains, we die. That's it. However, this goes against all the evidence we have considered so far about our quantum selves and their abilities to transcend time and space as we know it.

There is no doubt that the brain itself is an amazing computer which acts as a medium between the quantum self - which is the actual source of consciousness - and the physical self - and transmits information from the quantum field via the quantum computer of a brain we possess - enabling us to participate and negotiate the physical world courtesy of the five senses. When we die, the soul/quantum self leaves the body.

You have a 'ghost' inside you right now. So do I. This 'ghost' is the quantum field of energy that makes up your totality and which created you and to which you will return upon death. You will however, still contain the record of your individual personality and existence within that quantum field, within your soul.

Your life is recorded on the quantum level, at the soul level as a form of data. Remember those who had escaped sudden death and said 'my whole life flashed before me in a matter of seconds'? It's all being recorded folks. That's right. Your whole life - and mine obviously - is going on record. Why? No-one records data in this detail unless it's going to be used or useful. Perhaps one day we will have to give the DVD of our lives to the Supreme Creator and answer for our actions? Just saying!

The Theosophists Rudolph Steiner and Edgar Cayce believed that all our lives and the lives of all beings, are accounted for in full, in the Akashic records, which exists on another plane and which are a record of all universal events, thoughts, words, emotions, past, present and future. A Library of Souls if you will. Whilst there is no scientific evidence for this of course, it would make sense wouldn't it? Keeping an inventory of all that goes on in the Universe? Many people who have had near death experiences report seeing their whole life flash before them in seconds. This often happens during a sudden life or death crisis.

On May 22nd 2020, after a failed second attempt at landing, a routine Pakistan International Airways flight from Lahor to Karachi crashed into a residential district, killing all but two of the 99 people on board as well as many on the streets below. One of the only two passengers who miraculously survived the flight was bank executive Zafar Masood. Zafar recounted his experience.'I was moved seats at the last minute for some reason. I normally book to sit elsewhere. The flight seemed normal at first but you could tell something was wrong. When the plane finally crashed, I saw my whole life till this very moment flash before my eyes - as if on a canvas. I felt that this was it, I was going to die. Then suddenly, I heard a voice and it told me 'don't be frightened, it's not your time.' Suddenly, the seat where he had been sitting fell free from the plane, landed on a roof, and then fell again onto the street. He survived.

Who was the voice who told him 'it's not your time?' It is not uncommon in such incidents for people to hear those words. Many people who have had near death experiences have gone 'to the light' only to be sent back with the very same words 'it's not your time'. Whose voice is it? Is it God, a Guardian Angel, a spirit guide? It also indicates to me, at least, that there appears to be a written destiny to all things. Known only to a higher power. Our souls come into this world the same way and go out of it the same way.

It is not just the near-death experience where it is possible to access your past in full detail either. Any competent hypnotherapist could say to any of us right now (under hypnosis of course) 'Tell me what you were doing 5 days after your 5th birthday' and you will be able to give a complete record in perfect detail.

However, seeing is believing, and if you have never seen a 'ghost' yourself I won't blame you for being cautious in your view. However, ghost sightings - which were once for the eyes of the beholder only - are now available for all to see, either on the plethora of TV ghost hunting shows or on social media sites. With the advent of the mobile phone camera, there is no shortage of genuine footage shot by people from all over the world with seemingly valid visual evidence that you simply cannot ignore.

However, let's get a few ghost facts to start with. Firstly, a ghost is the spirit of a person who once lived on Earth who for some reason is either earth bound or has returned to contact a loved one. Secondly, ghosts don't talk. They have no physical organs of speech. They can only communicate telepathically, although you will hear it as sound but only in your own mind. Thirdly, ghosts do not walk. They have no physical body. They are not bound by gravity. They travel by floating, levitating or flying.

Your 'ghost' also leaves your body every night when you sleep, attached by your very own silver cord. We all do this every night! I personally believe dreams are actually the recollections of our activities on the 'astral plane' when our quantum selves are out of the body. This is also how a 'living' ghost can be seen at distant locations when out of the body during astral projection.

If you suffer from sleep paralysis, this is often when the astral body (quantum self) is just about to re-enter the body and you become conscious of it. Other effects of semi-conscious awareness of 're-entry' into the body, are landing back in your body with a sudden bump. However, when we die, the silver umbilical cord that connects the quantum self (soul) to the physical body is finally broken and the soul is released.

Have you ever seen the film 'Ghost'? In it, Sam - played by the late Patrick Swayze - is killed in a bungled attempt to rob him of a computer code that is the key to a fortune. At the moment of death, however, from Sam's body his 'ghost' emerges. He looks just like Sam, can walk through walls, project himself wherever he wants to be, communicate with the living - except that he is physically 'dead'. He subsequently learns how to affect matter, taught by a fellow ghost he meets on a subway train. Wisely, long dead, the subway ghost tells him: 'You don't have a body now, everything is up here (pointing to his head). It's all in the Mind now.'

Have you ever seen a ghost?

At the University of Virginia in Charlottesville in the USA, Professor Ian Stevenson has been studying reports of ghost sightings in Britain and the US and says that the people who see ghosts had normal healthy minds and believes that the reports are too frequent to be dismissed. In the UK upwards of 35% of people believe in ghosts, in the USA it's nearer 45%. Many people reported actually seeing ghosts and even communicating with them in some fashion.

What is a ghost? And can we prove their existence?

A ghost, quite literally, is the soul of the individual, the ball of transparent subatomic matter or pure consciousness, which retains the individual memory and personality of the individual life that leaves the body at physical death. The spirit, or soul, is the conscious ball of divine light and energy that lives within you - the quantum self! Because of your soul, your quantum self, you do not die when your present body 'dies'.

'Ghosts' are the souls of people who have died who for one reason or another have not yet left the earth plane. Due to attachment to a person, place, or event, we may be drawn back to the Earth after our physical 'deaths'. Often this is because the soul either wishes to say goodbye, assure a loved one they still actually exist - albeit without the physical self - or return with a message after death to friends or loved ones. In the worst case scenario, a soul can become earthbound, unable perhaps to leave the scene of some previous life, in most cases because of great tragedy or trauma.

One of the most amazing pieces of ghost footage that I have personally ever seen, for reasons I shall explain shortly, was filmed in the USA. The 'ghost hunters' in question were filming at the site of a motor accident where a young man had been killed some years previously, and which was reputedly said to be haunted by his spirit.

The team had their cameras running one night, when suddenly amongst the trees they could see a ball of light, an orb of pure white light moving about six feet above the ground from left to right of the frame. However, what happened next was exceptional in terms of visual evidence.

As the orb was moving from left to right it suddenly started to change, and the orb then began to slowly develop into the form of a man walking along - albeit semi-formed and transparent in nature, that seemed to emanate from the orb and slowly condensed into a physical form.

The ball of light, the orb began to condense and form a physical body. News Flash! This is the same way the whole universe comes into being. 'As above, so below', the creation of an individual mirrors the whole of the creation of the entire universe. That is, light being slowed down in vibration to become matter.

I cannot stress this enough. Each of us is a miniature replica of the whole Universe! We start as light, or pure energy, and this energy is slowed down in vibration to become matter.

Now, reverse the process. When we 'leave' the body, we firstly appear in our astral form, identical to the physical body but made of a different type of translucent energy, then having no further need of the 'identity' would then revert to a ball of light! Reverse again. Material forms come from Light (orb, quantum self) condense to form an astral body, which then condenses further to produce what we consider a solid body. So, we go from energy, to matter and from matter back to energy in an undoubted endless cycle.

If you have seen a 'ghost', you may see them in one of these three states:

1) a pure orb of light.

2) A semi-manifestation where the ghost is recognisable as the person in question, but is seen as transparent, or

3) a full body manifestation where the person fully manifests and looks solid.

These are the three stages by which we all manifest. I repeat, each of us is manifested in the same way the whole universe is. Light condensing to matter. At death, we go back to the light, to the quantum matrix if you will. Unsettled souls however, may cling to the Earth, unable to move on'. These are the spirits, or ghosts, of which haunted house stories are made, and there are many stories of which you have heard. Here is one I heard first hand.

Sometime ago, whilst visiting Glastonbury for the weekend, I met an American tourist, who had just 'done' Stonehenge and was visiting Glastonbury as part of his itinerary. 'Where are you going to stay', I asked, making polite conversation. 'The George and Pilgrim', he replied. 'The George and Pilgrim'? (Had I heard right?) 'Yes, 'The George and Pilgrim.' 'Did you know that it's supposed to be haunted?' 'No.'

As I was already aware of its history I declined the immediate possibility of some refreshment in the bar at the George and Pilgrim with my American friend, as I was anyway due elsewhere. However, I agreed to meet him again the next day.

'Hi! I called out, as I approached this old and enigmatic building that had once been part of the complex of buildings that had surrounded Glastonbury Abbey. Built around 1200, the Abbey was virtually destroyed after Henry VIII called for the dissolution of the Monasteries around 1530. The George and Pilgrim was once part of the original Abbey complex and was formerly used as administrative buildings, and later as monks' cells.

'Well, did anything happen last night?' I continued, 'you see any ghosts?' I wasn't quite prepared for the reply I got. My new friend then proceeded to tell me of the previous nights events:

'When I arrived at the hotel, they took me up to my room on the second floor, which was obviously locked. The receptionist then unlocked the door in front of me. When she opened the door, however, the whole room had been smashed to pieces. The mirrors were off the walls, the plugs were broken and there was furniture and debris everywhere. She (the receptionist) was really shocked and flustered and quickly took me away.'

Many of the staff have also had strange experiences at The George & Pilgrim with poltergeist activity. A Poltergeist is usually a very unhappy and angry ghost.

The hotel staff confirmed the room had been locked since it was last cleaned, and rushed my friend away from the room, locking it again as they went. However, since the rest of the rooms were fully booked, they eventually had to put him in the room directly next door. Unfortunately, he didn't get much sleep. 'The noise in that room did not stop all night' he told me 'I could hear strange noises and constant shuffling and moving about and yet I knew for certain that no-one was in there.'

I can only imagine that an Abbot or monk who once lived in that room was overcome with grief and anger when Henry VIII confiscated his precious Abbey. Or perhaps it is the Ghost of the last Abbot of Glastonbury Abbey who was eventually hanged on Glastonbury Tor for refusing to conform to King Henry's wishes.

I felt rather magnetised to The George and Pilgrim myself, for some reason I could not explain, and I visited it at least three times again later that day. On the third occasion I pulled up enough courage to ask to see the part of the building my friend had stayed in. I obtained a key for a room on the feeble pretext of wanting to stay the night (I think not!) and climbed the tiny stone spiral staircase, which took me up to the first floor. I unlocked the door and had a quick look round the room. I then made my way surreptitiously onto the landing of the second floor where all the commotion had occurred the night before. I didn't stop long however. Although it was daylight at the time, and I neither saw nor heard anything unusual, there was a distinctly strange and tense atmosphere in that part of the building. Shivers ran down my spine. Not wishing to tempt my fate, I left my research at that, and walked back out into the sunshine.

Stories of these sorts of incidents abound, especially in older buildings, and seem to occur when earth bound spirits have had traumatic lives, or experiences in that particular location. Perhaps their traumas were so great that they are still reliving it, unable to move on.

Yvette Fielding and Carl Beattie set up their own production company 'Antix Productions' specifically to investigate and film paranormal activity in supposedly haunted locations. Their highly popular series 'Most Haunted' caught the public imagination, and a spin off programme 'Ghost-hunting with Yvette Fielding' was later screened featuring special guest stars Louis Walsh and his band, Boyzone, who were taken to Edinburgh to visit the extremely eerie underground vaults in and around St Mary's Close, a subterranean city built into the rock, and once used to house criminals and plague victims.

Whilst sceptical about the paranormal to start with, each of the band had strange poltergeist experiences of their own to report - hearing strange voices, stones being thrown out of nowhere, tapping noises, footsteps and so on - but one unusual occurrence was caught live on film when one of the band members filmed a heavy metal ring attached to the dungeon wall (which had once been used to chain criminals to) starting to swing from left to right entirely of its own accord! I saw the footage myself. There was no one else in the room.

One of the most well documented modern ghost sightings, however, happened in America, after the crash of an Eastern Airlines plane into the Florida Everglades. Along with many passengers, the Pilot and Co-pilot were both killed, having fought valiantly to avert certain disaster. Subsequent to the crash, salvageable parts of the wrecked plane were used in other Eastern Airline planes and it was on those planes that the 'ghosts' of both the pilot and co-pilot were reported with amazing frequency.

They would suddenly appear in the cockpit, hold, or sit in passenger seats much to the alarm of the passengers and crew. On two separate occasions the ghosts of the two crewmen who had lost their lives, appeared to other pilots to warn them of engine fires, and thus averted two potential disasters.

Reports of these sightings became so numerous and the rumours on Eastern Airlines so rife, that it was fully researched and written about extensively in the book 'The Ghost of Flight 401' which was subsequently made into a feature length film. Eventually a medium was called in to exorcise the souls of the pilots who still felt responsible for the passengers and crews of other planes, and the sightings eventually stopped.

Families too, see or dream of the ghosts of their loved ones in the days and years after their deaths. Jenny Freilich is a free-lance Television Producer who was extremely sceptical about the paranormal until her own experience shortly after her father's death. Understandably upset just two days after he died, Jenny woke up in the middle of the night, when something in the corner of the room caught her eye. She looked up, and there saw the 'ghost' of her father whom she identified without doubt, as he was pulling funny faces at her that he, and only he, had done when she was a child when he was trying to cheer her up! Jenny told me that he did not look solid, but rather looked 'transparent', as though he were made of light.

Jenny later visited a medium who accurately described her father in detail and who, Jenny says, gave her a message from him, detailing information she (the medium) could not have possibly otherwise known about.

At this point we should mention Orbs! Are Orbs the spirits of those passed on? Yes, they are! The ball of light is the quantum self, the eternal energy that exists within us all. This orb of light is the essence of the self at the quantum level. It is where the identity of the self and its personality and history resides. However, If the spirit wishes to manifest, it will then proceed to create a form of energy which then condenses into material form.

You may see a ghost purely as an orb of light, or secondly, a ghost who is recognisable as your loved one or relative, (although seemingly transparent), or lastly as a fully manifested physical form.

As we saw earlier, the spirit body - being of a finer atomic substance to the physical body - is literally able to pass through what we consider to be solid objects. The slower an atom vibrates, the more solid it appears to be. The atoms of the etheric, or soul body, however, vibrate at a much faster rate, and are thus enabled to pass through the slower moving atoms.

Thus ghosts are enabled to walk through apparently solid objects and many people have witnessed them doing that in exactly the same way that people appear to do during the out-of-body experience.

The difference is, that in the case of the out-of-body experience the soul is still connected to the physical body by an ectoplasmic cord, and will return to its body after its journey. The soul of the deceased however, has already cut the cord and will not re-enter that same body again.

Mediums

'O descendant of Bharata, he who dwells in the body is eternal and can never be slain. Therefore you need not grieve for any creature'.

Bhagavad Gita 2.30

Although some people may experience seeing a departed soul, like Sam in 'Ghost', if the living are not sensitive enough to detect the presence of the person concerned for themselves, they may try to contact them through a 'medium', that is, one who mediates, or is the medium, by which the 'dead' can communicate with the 'living'.

Of course, this is a prime target of ridicule for the scientific and medical establishment and sceptics in general who think it all rather airy-fairy, being more likely a form of 'cold reading', or simple trickery.

The following story however, proves beyond a shadow of a doubt that there is more to mediumship than trickery, when a medium discovers things about the deceased that no one knew - and can prove it!

James Van Praagh is a well-known American medium and author, and in his book 'Talking to Heaven' he describes the dramatic case history of a young man who had apparently committed suicide.

The boy's parents went to James on the recommendation of friends because they were so distraught after their son's death. According to James, they seemed quite sceptical at first, and unsure of getting involved in Spiritualism. During their first session however, James began picking up messages from 'the other side'.

James Van Praagh: 'There is a young man here who passed over very quickly. You've been asking for him.'

'Yes', they replied. The couple both began to get visibly upset. Their son had apparently committed suicide one year previously. James correctly identified the cause of death - he had been shot with a gun - and the location of the body at death. He also identified a drug problem their son had had. 'Yes, we found that out afterwards'.

James: 'Your son is very strong, He is yelling - it was Ronnie! Who is Ronnie?'

'Ronnie was a friend of his', they replied.

James van Praagh then began to get a message about a gold watch.

James: 'His watch. He's talking about his watch.' (The boy's parents explained that they couldn't find the watch after their son had died. 'We looked everywhere.')

James: 'Your son gave it to Ronnie for payment. Ronnie was angry. Do you know if there was a fight before your son died?'

'No'.

James: 'Steven is screaming at me, I didn't kill myself. It was Ronnie. Ronnie did it to me. I didn't kill myself!'

No one could believe what they had just heard. Steven had been found dead, alone in his room, with a gun at his side, an assumed suicide!

During the rest of the session James van Praagh was able to describe many other details of the incident and the location of Steven's watch, which eventually lead to the arrest of Ronnie, who was finally caught with a kilo of heroin and Stevens gold watch hidden behind the wall of his (Ronnie's) garage.

After being questioned by the police, Ronnie eventually admitted that he had shot Steven in an argument over money and is now serving a life sentence in a state penitentiary.

Although I could give you endless examples of the work of different mediums, they are subjective and the example of James Van Pragh is provable. However, the question you might rightly ask here, is 'how on earth can you talk to people who are dead'? The answer, when you know it, is remarkably simple. Telepathy. Telepathy works regardless of time or space, or whether or not the consciousness, the quantum self, is inhabiting a body or not.

We all have telepathic skills, but for many they remain dormant. Nonetheless, we all have them because we are all connected to the quantum web. Whether you realise it or not, you have the same abilities.

Telepathy works however distant one is in time and space. You could be ten feet from the source, 10 miles, 10 million miles - telepathy works regardless of distance. To the cosmic web, with the concept of quantum entanglement, it doesn't matter. It's all connected. When someone deceases to be in their body, their consciousness still exists and the method of contacting those on the other side is no different. Its telepathy. It works regardless of whether you are in your body or not, whether you are in this dimension, or the next, because consciousness is not in the body and is inexorably linked to the quantum world through entanglement.

Of course, mediumship has been the victim of many sceptical critics. First popularised in Victorian Britain, at a time when writers and artists were investigating alternative realities, including both the afterlife and psychoactive substances generally found in cough medicine, spiritualism and seances became extremely popular parlour friendly pursuits for both rich and poor alike. Many stories then abounded about fake mediums using trickery to extort money from their vulnerable victims, conjuring up fake ectoplasm and controlling the sitting through dubious means - and this may have been the case in some instances of fraud. Unfortunately, the propaganda left over from that time still lingers, and mediums are still often considered to be fakes or frauds. However, a simple understanding of the mechanics of the matter may help allay some fears. There is no trickery to genuine mediumship and there is a working code with mediums today which prohibits such fakery.

Given the success of James Van Praagh helping solve an otherwise unsolvable murder, perhaps it's time to take these skills more seriously. Perhaps mediums could be invaluable in solving other unsolved murder cases and indeed, are often called in to help the police with unsolved crimes?

Why do ordinary people visit mediums? Our desire to contact the ones we love who have left this world is natural. Indeed, in many cases where relationships have been difficult in life, the 'deceased' is able to return briefly, maybe to set a loved one's mind at rest, to apologise or to set the record straight.

One of the most unusual cases of mediumship however concerns Air Chief Marshal Sir Hugh Dowding who invented and organised the radar air defence system and presided over the Battle of Britain during WW2. Ten years prior to the war, however, having lost his wife shortly after they married, he turned to spiritualism and visited many mediums in an effort to contact her. Apparently with some success.

That was not the end of it however, as later on during the war, he himself apparently developed mediumistic abilities. Dowding had become close to many of his pilots - who he felt immensely responsible for - and when some of them were eventually killed in action he claimed that these young airmen came to him as ghosts. Many of them did not realise they were dead and Sir Hugh claimed that he was able to comfort them and guide them onwards. When he saw them he described them as 'spirits who flew fighters from mountain highways made of light'.

This did not go down too well with his colleagues of course, and he was ousted from his post in 1940. He obviously had mediumistic abilities but unfortunately, had to suffer the ridicule that comes with it. Sir Hugh was also a member of the Theosophists Society and was a firm believer in reincarnation.

The medium is usually clairvoyant, or clairaudient, (like Odessa May in Ghost) but also able to contact spirits who are 'on the other side'. To a medium, they may hear, or see, messages from beyond, and sometimes even 'feel' the physical afflictions the deceased may have departed with.

In many cases, the deceased return looking as they did when they were young and healthy, because don't forget, on the other side - time as we know it simply does not exist!

Spirit Guides

'You have listened well to my teachings. Now I must return to the spirit world'.
                    Buffalo Calf Woman. Sioux Indian Spirit Guide.

Have you ever felt that someone was watching over you, or that sometimes you seemed to be guided towards certain things you were not consciously aware of?

Do you perhaps have a favourite aunt, uncle or

grandparent who has died, whom you felt particularly close to, who may still protect and guide you from 'the other side'?

A growing number of people, even in today's modern materialistic and technological age, claim to have spirit guides, or to have had close encounters with their own guardian Angels.

Actor William Roach, who played Ken Barlow in the British TV Soap drama, Coronation Street, believes he saw an Angel shortly after the death of his 18th month old daughter, Edwina. At the time, his grief was so profound that he became physically ill and could neither eat nor sleep. On the morning before her funeral, however, he awoke to an incredible sight.

'There was this glorious round glowing light and in the middle of the light, there was Edwina's little face smiling up at me. At the same time as I saw this - and there is no question that I did, it was not a dream, it was not imagination, it was real - I felt the greatest feeling of love I had ever known.' From that point on, his grief lifted. 'When I saw her, I knew an Angelic being was helping her appear like that. I now know of the existence of Angels. This has helped me cope. I don't need any further proof. I am aware of their presence and how they help us'.

Diane Davey - a Buddhist - didn't believe in Angels until her husband was in the final stages of terminal cancer. 'I went to check on him, as I always did, and saw he was asleep in the chair. But to my surprise, I saw a huge Angel standing behind him - at least six or seven feet tall. She was a classic angel, but it was as though she had rays of golden light flowing out of her. There was so much love in that room, I wasn't scared'.

Diane saw the Angel again on several occasions until the final occasion, on her husband Robert's death. 'When he died, his bed was surrounded by Angels. After he died, I felt he was safe being taken care of. I once thought Angels were only for Christians, but now I have seen on, I believe in them too'.

Is the desire to see, or believe in Angels a reaction to grief or death, or do these heavenly beings really exist? It would appear from the descriptions thus far, that Angels are somehow made of some radiant form of Light and are not subject to the laws of time, space or gravity. We have already seen that we are composed of this same radiant form of light but we know we have physical bodies in which this light exists. Because of our material bodies, we can only operate on the material plane. Angels, however, are beings of Light from a higher spiritual dimension that can manifest to us at will.

How can Angels exist in other dimensions?

We have already seen that the idea of a ten dimensional universe exists with both scientific string theory and Kabbalistic teaching. Both these systems remind us that there are other dimensions, other planes of existence other than our own. Perhaps it is one of these dimensions on which spirits and angels can exist. We know the highest dimension consists of pure divine energy, consciousness or light - so could it be that Angels really are from Heaven - the source of Light - the highest dimension in the whole of creation?

Children, of course, can also see Angels!

When Sonia Lynch's five-year-old daughter Gabriella contracted cancer, she too began to see Angelic visitors on a regular basis. Gabriella told her mother that she could see beautiful angels who looked like 'sparkly coloured lights' and said 'Open your eyes mummy, they're everywhere - I can't believe you can't see them!' At the end, Sonia believes the angels helped Gabriella transcend the material plane and go on to the next level.

However, Sonia was to see her daughter one more time. Shortly after Gabriella's death she woke up one night and saw a 'white light' in the form of Gabriella sitting with the new baby in the baby's cot. Sonia now believes Gabriellia is a spirit guide to the new baby. 'It makes me feel wonderful to know that she's floating and flying around and can be anywhere anytime she wants, and can be there when somebody needs her'.

What is a spirit guide? A spirit guide is quite simply, someone who guides and protects us from the spiritual dimension.

We have seen that telepathic communication is possible during life from aura to aura or soul to soul. This communication is still possible after the physical death of one or the other of the parties. Physical death may have occurred, but the soul is still intact, and can communicate to 'living' relatives or friends if there is a strong enough morphic connection between them.

Many people who have lost loved ones often claim to feel their presence on numerous occasions after physical death itself. How is this possible? The soul is a ball of light that is no longer subject to the restrictions of the material dimension and the space-time reality we are familiar with. That is why when people see 'ghosts' they see the light without the body - the spectre or soul of the person concerned.

Now, do you remember the story of Chris Robinson that we looked at earlier, and do you remember it was his dead grandmother who first visited him in a dream that activated his psychic activities? When friends or relatives 'pass over' they often come back to help or guide us, even though we are not aware of them. They may talk to us in dreams, or be there to guide us through our daily lives. One especially thoughtful brother who had just passed on was kind enough to return in a dream and give his sister the winning lottery numbers. She subsequently won millions of pounds!

This is rare, however, as most guides are here to help us with the spiritual aspects of our lives, although occasionally their help will be of a more practical bent. In some cases, spirits of great doctors or writers, may visit the Earth Plane to assist someone in need of their expertise, and many of today's healers claim to have medical experts as their guides.

Many guides who seem to appear today are from older and more spiritual cultures. There are many people who claim to be guided by American Indians - or Egyptian priests and priestesses. This is because they are here to help raise the spiritual vibration of the planet at this time. They are helping us to remember our ancient mystical wisdoms.

Many mediums, healers, psychics and clairvoyants also claim to have Spirit Guides, that is, a soul on the other side, who retains an interest in their lives and helps them in their work.

However, one of the most fascinating cases of apparent Spirit Guide communication I have ever heard, concerned the case of ex-Sunday Times journalist, Hazel Courteney. In her book 'Divine Intervention', Hazel tells the story of her own personal encounter with the spirit world when, shortly after Princess Diana's death she found that she was able to communicate with her spirit. Hazel describes in detail finding herself elevated to a spiritual dimension, guided and instructed by Diana from the spirit world. She suddenly found herself able to heal, which she claims she did by emitting a beam of light from her eyes.

Although the idea of light beams shooting out of the eyes may sound rather like science fiction or at the very least the results of hallucination or psychosis, research at the Ukhtomskii Military Institute in Leningrad conducted by Genaday Sergeyev, a top neurophysiologist, confirmed that a light beam, or light force, which is emitted from our eyes, actually existed! He subsequently named it Odyl.

Sergeyev used Kirlian photographic techniques to capture the image of the aura of Nina Kulagina, a psychic sensitive who could demonstrate the moving of objects without touching them.

His photographs revealed that while Kulagina performed these psychokinetic feats, the bioplasmic field, or aura, around her body expanded and pulsed rhythmically as a ray or luminescence seemed to shoot out from her eyes!

Hazel also found herself producing ash from her hands - in the same way that some Indian masters including Sai Baba are said to be able to do. Hazel had literally found herself in between worlds, the world of spirit and the world of matter. She saw parts of her material body disintegrating and reintegrating before her eyes, and somehow found herself able to materialise physical objects at will! The experience however, nearly cost Hazel her life.

After Diana's death, literally thousands of people reported seeing her 'ghost' in various locations, or feeling her presence. Did Diana come back with a message of hope for humanity that she did via Hazel, whom she had met on several occasions before her death? If so, we can rest assured - death is not the end and when our time comes, the Angels will be waiting.

Chapter Six.
IMMORTAL MIND

The Law of Cause and Effect.

'For every little Action - There's a Reaction'.
Bob Marley.

Everything you and I think, say or do, has a knock on effect on the rest of our lives, and the universe in which we live. Sir Issac Newton defined this law, as The Law of Cause and Effect, otherwise known as the 3rd Law of Motion.

Every action, however small, has a reaction that has a knock on effect on all of life. In the same way, our thoughts and actions are also subject to the law of cause and effect.

The Law of Cause and Effect, also known as the Law of Karma, states that every cause has its effect and every effect has its cause. Every action or thought has an equal and opposite effect. The nature of the effect depends upon the nature of the cause. In other words, 'we reap what we sow.'

With every thought or action, we set into motion an invisible series of causes and subsequent effects that will vibrate from the mental plane imbuing the entire cellular structure of our bodies and then manifest onto the physical plane, and ultimately into the entire Cosmos. Eventually, these actions return to us as reactions, in other words, 'What goes around, comes around.'

In applying the Law of Cause and Effect, providing we have had good intentions in our actions, we are ultimately assured of a beneficial result.

This law like all other universal laws is unchanging, unbiased and immutable. Universal Laws are at work everywhere, all the time and without exception. The Law of

Cause and Effect is a literal law both physical and spiritual which is a direct result of our own thoughts and actions. The Law of Cause and Effect has both a scientific and metaphysical reality.

How does cause and effect work?

Let us take a simple example and build an imaginary line of dominos. Then let us knock the first one down. What will happen? Exactly. Even though we have not touched anything except the first domino, the knock on effect will be that it will knock all the other dominos over, culminating in the last one in the line being knocked over.

Although no one actually touched the last domino, or any of the other dominos, save the first, it was the initial action of knocking down domino number one that has brought the inevitable reaction. The Last domino was not touched at all - it was the action of the first domino falling, which had the inevitable effect of knocking all the other dominos over. Domino number one, and the last domino to fall had no direct relationship with each other. It was entirely as a result of the number one domino becoming activated that the chain of events finally culminated in the last domino reacting.

Simply put, this is how cause and effect works. Every thought or action you and I take has a knock on effect on the rest of our lives, and the rest of the universe. Now, let us imagine that our thoughts are like a circle of dominos. Thoughts go round the universe and come back to you as sure as day follows night and night follows day.

Doesn't it make sense then to make sure that your thoughts and actions are positive and not negative. After all, you're the one at both the beginning and the end of the line!

The Law of Karma

'As a man soweth, so must he reap'.
The New Testament. (Galatians 6:7)

What is the Law of Karma?

The law of karma is the law of cause and effect operating in our lives. Karma is the result of our own actions, all of which have opposite and complementary reactions, which will eventually return to us in like form. This is in accordance with the Law of Attraction. We attract to ourselves that which we have already created as a personal reality. Therefore, if we send out positive thoughts and create positive action, we get positive results. If we send out negative thoughts and create negative actions, we get negative results. Like the circle of dominoes, once we have created a thought and set in motion its resultant actions, it will have a knock on effect on the entire universe, eventually returning to us in equal measure.

However, all our karma cannot possibly be fulfilled in one lifetime, and it is through the concept of reincarnation that the soul is enabled to return to fulfil its own karmic destiny. The soul cannot become perfected in one lifetime alone. It needs countless lifetimes to reach enlightenment and to overcome all attachment to earthly desires. In essence, to become enlightened, the soul must become aware of its perfect spiritual divinity. It must become pure.

Only in this way, it is said, can we end the countless cycle of births and deaths. According to the law of karma and reincarnation, so long as we seek gratification of our desires in the material world, we shall be creating new karma for ourselves, whether good or bad and we shall have to return to fulfil it. For good or ill. The concept of spiritual karma, or divine judgement for our actions, is clearly inherent in all the world's religions and all, without exception, stress the importance of right living and the consequences of doing harm. It is the very basis of our moral laws.

Many great thinkers throughout the ages have also adhered to the ancient concepts of karma and reincarnation. The Greek sage, philosopher and mathematician Pythagoras was once overheard berating a passing stranger 'do not beat that dog. He is an old friend of mine incarnated again. I recognise his voice!'

Although to many, this may sound like a rather comical anecdote, the concept of the soul reincarnating in a variety of forms and guises to further its karmic learning is accepted in many religions and philosophies worldwide. In the Eastern religions in particular, there is a total acceptance that our individual fates are entirely of our own making and that our current lives are the logical results of our own karma - the fruits of our own past action.

Despite its increasing popularity and acceptance in the West there are still many who find the concepts of karma and reincarnation rather alien and in some cases, rather offensive to their understanding. Regardless of our sensitivities and preferences, however, the law of karma, or the law of cause and effect, exists as both a physical and spiritual reality. They are unbroken universal laws.

The theory of karma and reincarnation, or the law of cause and effect, is hard to understand at the best of times - even for the most enlightened amongst us - and even harder to explain! So much so, that I'm sure even the Dalai Lama has trouble with it from time to time!

Although the idea of karma, that is paying for our actions and our sins may be an idea abhorrent to many Westerners, who prefer to think that our individual lives and fates are no more than the fortunate or unfortunate throw of a dice, I personally think the idea of a random destiny and NOT paying for our sins to be an even more abhorrent idea! If everything was entirely random, and none of us ever had to pay for anything we did - the concepts of order and justice would die - instantly!

Do we 'pay for our sins'? It tells us as much in the Bible. It tells us so in The Quran. It also tells us clearly in the ancient primal and nature religions of our ancestors, and in the Eastern philosophical texts of Buddhism and Hinduism.

In essence, the theory of karma and reincarnation states

that all of our actions, whether positive or negative, will have ongoing reactions, and until we free ourselves from attachment to earthly desires - therefore ceasing to create either action or reaction - and perfected and purified our souls, we shall not escape the necessity of endless rounds of incarnations, births and deaths that we must suffer until we reach absolute perfection! Until we become enlightened and purified.

Is the idea of Karma and Reincarnation inherent in Judaism and Christianity or Islam? The idea of reincarnation has certainly been a part of Judaism since the time of Christ, and if Christ had travelled in the East he would certainly have learnt about these laws from the Indian Yogis and Sages.

However, the concepts of karma and reincarnation have been suppressed in mainstream male dominated religion, and so far as Christianity is concerned, the idea of reincarnation was deliberately omitted from the official tenants of Christianity at the Council of Nicea. Basically, it was binned! Any reference to reincarnation in the original scriptures will now do doubt be in the vaults of the Vatican.

Although excluded from the main teachings of Christianity, Judaism and Islam, each has its own mystical sects who accept the theory of karma and reincarnation. The Hasidic Jews in particular, retain this ancient belief and regularly pray for those who have 'sinned against me in this, or in any other life'.

Paul certainly seems to confirm Christ's teachings on the effects of the law of cause and effect, the law of Karma when he says 'God will repay each one of us as his work's deserve...pain and suffering will come to every human being who employs himself in evil...Renown, honour and peace will come to everyone who does good...God has no favourites.'

In his letter to the Galatians Paul says: 'Every man shall bear his own burdens...be not deceived... God is not mocked - for whatsoever a man soweth, that shall he also reap. For he that soweth to his flesh shall of the flesh reap corruption:

but he that soweth to the Spirit shall of the Spirit reap everlasting life.' Paul continues: 'Every man shall receive his own reward according to his own labour'.

Is this a reference to the Law of Karma? It would certainly seem so.

But Karma is incomplete without reincarnation because one simply cannot fulfil all one's Karma in one lifetime! The thought impulses that have become action will continue forever to affect both energy and matter. If the action was negative, we shall reap negative rewards, and vice versa. We, as energy forms, are not restricted to one life on the material plane. We are energy and we will re-create a new form for ourselves when we continue our energy journey of cause and effect and recreate an appropriate new life for ourselves.

The theory of reincarnation cannot be separated from the law of karma. Because of the eternal nature of cause and effect, the reactions to our actions continue until we have achieved spiritual purity and perfect equilibrium! The law of cause and effect operating as spiritual karma is an absolute mystical, spiritual, physical and scientific law and there are no exceptions to this law! Whatever we do comes back to us. It is the spiritual law of cause and effect in action. It is the greatest and most natural of moral laws. Ever.

Furthermore, the concept of judgement and some form of afterlife is inherent in all our primal religions and referred to in many great monotheistic and eastern religious texts, the Bible and The Quran included.

In part two of The Quran, verses 26-7 it states: 'How do you disbelieve in God, seeing you were dead and he gave you life, then He shall make you dead, then He shall give you life, then unto Him you shall be returned. It is He who created for you all that is in the earth, then he lifted himself to heaven and levelled them seven heavens, and He has knowledge of everything'.

In virtually all religions, the concept of divine judgement for our sins is taken profoundly and seriously. Perhaps it is a lack of that belief that has allowed mankind to become so sinful!

After all, would Adolf Hitler have been so ready to condemn his eternal soul to life after life of misery in return for his dreams of World domination for instance, if knew would have to return to pay for his sins in full measure? For, 'What profiteth a man if he should gain the world, and lose his soul?'

Would we be so cruel to each other if we knew we would have to come back one day to pay for it all? Would we enslave a man when one day we knew we must return to be enslaved ourselves? Would we starve and beat each other, rob, murder rape and pillage, if we knew that one day, we would be called to account for our sins and suffer in equal measure?

The Law of Karma and Reincarnation is the greatest spiritual and moral law in the universe.

However, suffering for our own sins is only part of the law of karma, for we are also rewarded for good actions! Now, these cannot be hollow acts intended for our own benefit. That is not proper action. Proper action is when love and compassion determine all our actions.

Despite any handicaps we may have, physical, mental, emotional or spiritual, the beauty of the law of karma and reincarnation is that once you have learnt your lessons and begun to align with true spirituality and morality, your lot in life immediately begins to improve! So much so, that right actions will bring their own good rewards. This too, is carried from lifetime to lifetime.

The law of cause and effect, or the Law of Karma is a Universal Law and it is operating within the Universe at all times. The Law of Karma operates without preference or prejudice. It is our own selves who dictate our life journeys through our own thoughts and actions. However, it is not just action alone that brings good or bad karma.

Thought, as we saw earlier can also be felt and experienced as a force in its own right. Both positive and negative thoughts will have consequences. If you remember that thought waves are continually emitted from the self, via the aura, or electromagnetic force field, and if you consider that the Universe, being cyclical by nature, will eventually send those very thoughts and deeds back to you, you begin to understand something of the nature of karma. Karma is natural justice.

Some Karma is instant, some takes many lifetime's to fulfil. The racist murderers of Stephen Lawrence for instance, may have corrupted and escaped the due process of Earthly law, but there is no escape from the justice of Heavenly Laws. One day, they will, like the rest of us, be called to account for their deeds. They too will have to 'reap what they have sown'.

No one escapes universal law.

Now, far be it for me to judge anyone - that is for God alone - and in the New Testament it tells us: 'Judge not, lest ye be judged. For with what judgement ye judge ye shall be judged and with what measure ye mete it shall be measured to you again'. (Mathew 7:l-2) because I, like you are human, and being so, we do not know God's absolute will in the order of things. We only see things from our individual perspective.

Now, imagine that one individual lifetime was a single frame on a roll of film, each frame representing one individual life. We, as 'humans' are usually only conscious of one life at a time, and to all intents and purposes are unaware of the others that if they exist at all, exist within the past and future unconsciously.

Only by viewing the whole film can we understand the plot. Each life-time is like a single episode in our journey, each one linking each to the other in an endless cycle of incarnations, until the soul has reached perfection.

The Kabbalistic tree of life shows us how energy manifests from Spirit to Matter. But as matter we must return, matter to spirit. Like a river to the Ocean, our souls seek their source. But we must purify our souls and become fully enlightened before we can return. We are all here for this common purpose.

That is why we cannot judge others in an absolute sense - we simply do not know the whole story. We do not know the plot. In the words of Christ : 'Let him who is without sin caste the first stone'. (John 8:7) We cannot caste one stone. This is where forgiveness, humility and compassion become the highest and most noble of attributes.

When we forgive wrongdoers, we free ourselves, and when they eventually forgive themselves - and I say 'eventually' because this can sometimes take many life times of soul searching - they too, are free. However, there is no excuse or escape from the karma of sin. Destiny will punish us until we learn to face our own eternal spiritual divinity and we will not be free until such time.

Now, the Universe is - so far as we know - spherical in shape and because of this any wave, including our thought waves, which are sent out to the far reaches of the Universe, will eventually come back to where they started from. In the same way, for example, if you started walking around the world from Paris, going East, you would eventually return to Paris, coming from the West. Similarly, thoughts eventually come back to the sender.

Karma is the law of cyclic energy reacting and returning to itself. Thought waves will pass around the Universe and come back to you - sooner or later, just like the domino circle.

Now, earlier, we discussed the concept of destiny and free will, which will surely be a question on everyone's mind. So, do we have a written destiny, or do we have absolute free will? The answer to this controversial question is equally controversial, and this, I believe, is the answer - we have both freewill, and a written destiny. How can that be? How can we have both? To answer this question in the best possible way I can, let me draw you a simple analogy to explain this in the simplest terms.

You are waiting at a train station. You need to make a journey and you must now choose between, let us just say for now, train A, B, C or D. Each train will take a specific route on your journey to your predetermined destination, and you are entirely free to choose which journey you will make. However, once you have made the decision and boarded the train, you must follow the route the train takes implicitly. You must follow the route you have already chosen.

Now, let us just say that when you are at the train station, you are in your Pre-Birth Consciousness, if you like, in between lives, where your soul must decide what particular life journey it needs to make to continue its karmic lessons on Earth.

Once you have chosen, once you have determined which journey you need to make however, which train you will take, which lifetime your soul needs to continue it's karmic journey - it's individual energy cause and effect scenario - you must then incarnate back onto the material plane, and follow that route implicitly.

Once the spirit has decided - with its own free will - what type of life it must experience for it's soul development on the Earth plane to fulfil its karmic debts or reap its karmic rewards, the spirit then selects the parents and the circumstances of the time and place of its birth, in order to - how can I put this - the nearest I can describe it is - it's like trying to jump on a moving roundabout - the spirit must jump enter the earth plane at the exact moment, at the precise hour that the planetary bodies are forming specific energy vibrations which will offer a path to the evolving soul, which will reflect the type of Earth life which he or she must take for its further soul evolution.

Now, you cannot accept the theory of predetermination to any degree unless you understand something about Astrology. If you go back to the concept that planetary magnetic effects influence individual life forms, and are entirely predictable - we may begin to understand the universal process. At the Soul level, we choose our moment of birth to continue our karmic lessons. We determine our fate. We have chosen our destiny.

Destiny and free will are two sides of the same coin. We have free will to choose at the spirit level - at the pre-physical level before birth - but once we have chosen a life path, we must follow the path we have chosen implicitly when we incarnate onto the earth plane - as described by the planetary influences. In other words, the soul, having assessed it's next karmic journey on the pre-physical plane, must select the appropriate entry to the Earth plane to access planetary harmonics which will describe the point that the incarnating one has entered the Earth plane at, and will describe the life patterns determined by the free will of the 'incarnatee' which it has already, at the soul level, deemed best suited for it's individual spiritual evolution and development.

We have to learn soul values by direct experience. Compassion is the highest goal. Experience the greatest teacher. It is not some vengeful God who judges us, it is ourselves. Our conscience. The pure part of God, the Light, which exists within each of us all at the soul level, it is that which judges and witnesses all our actions. You may have committed your sin in private, in secret - but God was watching and so were you! You cannot escape your own cause and effect scenario. What you do and what you think will come back to you in like measure, for good, or ill. You have written your own destiny. You have created your vibrations. Now they will return to you. We have made our beds, now we must lie in them.

In the Quran, Verses 24 it is written: 'Give thou good tidings to those who believe and do deeds of righteousness, that for them await gardens underneath which rivers flow; whensoever they are provided with fruits therefore. As for the unbelievers, they say 'What did God desire by this for a similitude?' Thereby he leads many astray and thereby he guides many and thereby He leads none astray save the ungodly such as break the covenant of God. And those who do corruption - they shall be the losers.'

The idea of a pre-written destiny, or Karma, is also inherent in many early nature religions and African primal religions where it is said that each individual is given a particular destiny by the Creator before birth. In Ghana they believe that destiny is determined by the way in which the new being takes leave of God before birth.

The paired concepts of karma and reincarnation can be traced back hundreds of thousands of years to almost every ancient culture, and the most basic forms of belief in an afterlife can be traced back even further. Despite our modern materialistic and technological lifestyle and the unrelenting opposition from modern medics and scientists, these concepts have now again become readily accepted and integrated into today's modern psyche.

The soul it would appear, however, is on an ongoing journey towards enlightenment. As long as we crave, as long as we desire, then we shall all return time and time again to Earth until we relinquish all material desires. In Buddhist terms, it is 'Non-Action', compassion and detachment alone that can release us from material bondage.

No doubt this is why Buddhist monks lead the kind of lives they do. 'Life is suffering' according to Buddhist thought, and which of us can say life has not been without its pains and struggles. But these are the struggles of our own making. Yes, but by living more conscious and compassionate lives, and having reverence for all life, instead of pandering to our own egos, we can instantly progress, in spiritual terms, to better karma, and a little closer to spiritual nirvana, or heaven. This is the beauty of Karma. We all get another go.

Wishful thinking? I think not. The Law of Karma, the law of cause and effect is a scientifically provable physical law. It favours no man, it gives just measure to our own actions. The

Law of Karma is an absolute moral and absolutely just law. Whatever we do, we pay for. Whatever we do, will come back to us in like measure, sooner or later. In The Shvetashvatara Upanishad verses l- 7 it is written of our fates:

'What is the cause of the cosmos:
Is it Brahman (God)?
From where do we come? By what do we live?
Where shall we find peace at last?
What power governs the duality
Of pleasure and pain by which we are driven?
Time, nature, necessity, accident,
Elements, energy, intelligence -
None of these can be the First Cause.
They are effects, whose only purpose is
To help the self rise above pleasure and pain.
The world is the wheel of God,
Turning round and round
With all living creatures upon its rim.
The world is the river of God,
Flowing from him and flowing back to him.
On this ever-revolving wheel of being,
The individual self goes round and round
Through life after life, believing itself
To be a separate creature, until
It sees it's true identity with The Lord of Love
And attains immortality in the indivisible whole.
He is the eternal reality, sing the scriptures,
And the ground of existence.
Those who perceive of him in every creature
Merge in Him and are released
from the wheel of birth and Death.'

The Theory of Reincarnation

'There are two ways of passing from this world - one in light and one in darkness. When one passes in light, he does not come back, but when one passes in darkness, he returns.'
Bhagavad Gita. 8.25-26

'Reincarnation is real, as consciousness is contained in the universe after death. Reincarnation is possible as consciousness is simply energy which Is contained in our bodies and which is released after death and can find a new host', according to Dr Jim Tucker who spent 15 years researching over 2,500 children who claim to have been reincarnated.

One of those cases was of a young boy called James Leninger who was the son of a devout Christian couple in Louisiana. They had no prior cultural or spiritual knowledge of reincarnation so they were intrigued by what their son was experiencing which seemed to have no rational explanation.

The boy, who was only 2 at the time, had no information regarding WW2, although he was strangely obsessed with his toy planes. Eventually he started having nightmares where he described being in a plane crash. He described how he had been a pilot in a war, and that he had flown off a boat. When asked the name of the boat, he said 'The Natoma.' He said he had been shot down by the Japanese and killed at Iwo Jima.

After detailed research they found there was an aircraft carrier called the USS Natoma Bay and that it was stationed in the Pacific in WW2 and was, indeed, involved in the battle of Iwo Jima. Records also showed that there was a young man named James Huston whose plane crashed exactly how little James had described, hit by a Japanese shell, bursting into flames and crashing into the sea.

Furthermore, this is not the only case of childhood past life memory that can be verifiable. In almost every case of children remembering past lives, they have been able to give information not otherwise available to them regarding historical details of the time they once lived.

Another young boy named Cade was 3 years old when he began to have nightmares about falling from a tall building which had exploded. Cade went on in detail and remembered a violent death on 9/11 where he had to jump from one of the towers to escape the fire. 'The plane hit, got stuck in the building and I had to jump to escape the fire. I was still alive when I was falling.' At first, his mother thought he just had a vivid imagination, but he had no prior knowledge of 9.11 so couldn't understand how he knew all this.

He then started to tell her 'Cade is not my real name. My real name is (the name has been withheld by Cade's mother for privacy reasons and respect to the 'other family'). Cade's mother then researched the name Cade had given her, and found the name that Cade had said was previously his, actually matched records of that person's demise on 9.11 and in the same manner Cade had described. Cade's mother wanted to contact the family herself to share this with them, but what would she say 'your son did not die, he's just someone different now?'

Again, traumatic lives or deaths seem to be the main factor in so very many of these childhood memories. These stories, many historically verifiable, are now so commonplace, particularly in the USA which has no cultural or religious relationship to the concept of Reincarnation that it is really quite phenomenal. The list just goes on and on. Twenty years ago, any researcher would have been hard pressed to find any examples of past life recall, finding them occasionally under hypnosis in adults and often unexpectedly or by accident. Now, they are literally coming faster than I can type them!

Even more remarkably - if reincarnation was not remarkable enough - many of these children also remember being in 'Heaven' with 'God' and looking down and choosing their parents. These memories are not an isolated occurrence. Many children who remember past lives also remember being in their pre-birth consciousness before incarnating back to Earth.

Dr Tucker, a psychiatrist from the University of Virginia, says that reincarnation is possible, thanks to consciousness being energy on a quantum, subatomic level which is contained in our bodies, but not a part of them.

To substantiate these concepts, a new groundbreaking theory holds that quantum substances actually form the soul and whatsmore, are part of the fundamental structure of the universe. According to this idea, consciousness is a program for a quantum computer in the brain which can persist in the universe even after death.

Mathematician and cognitive scientist David Chalmers is one of many who also believes that consciousness may be fundamental in the universe, perhaps equal with matter and energy neither 'derivable from or reducible to' anything else, including the brain.

Proof of this theory is substantiated further by Dr Stuart Hameroff, Professor Emeritus at the Department of Anesthesiology and Psychology and Director of the Centre for Consciousness studies at the University of Arizona, who has advanced a quasi-religious theory. It is based on a quantum theory of consciousness that he and British physicist Sir Roger Penrose have developed, which postulates that the essence of our soul is contained inside structures called microtubules within brain cells. They have argued that our experience of consciousness is the result of quantum gravity effects in these microtubules, a theory which they dubbed 'orchestrated objective reduction. (Orch-OR).

Furthermore it is argued that our souls are more than

the interaction of neurons in the brain. They are in fact constructed from the very fabric of the universe and may have existed since the beginning of time. With these beliefs, Dr. Hameroff holds that in a near death experience the microtubules lose their quantum state, but the information with them is not destroyed but merely leaves the body and returns to the cosmos.

However, they have yet to quantify how these microtubules operate as an organised energy field, because it is obvious from the stories of out of body, near death experience and past life memory that the person involved, however many lives they have experienced, always identify as themselves, that is as if the 'soul' or self has a definite personal identity and memory regardless of its many different incarnations.

General George Patton, who successfully led the US Army to help secure victory in Europe in WW2, was himself, like Sir Hugh Dowding, a firm believer in reincarnation. George Patton claimed he had been Napoleon in a previous life, and a soldier in many other lives. Would this explain his exceptional skill as a warrior? He once wrote poetically of his experiences of reincarnation. This is an extract from it:

'So as through a glass and darkly, the age long strife I see, where I fought in many guises, many names, but always me.
So forever in the future, shall I battle as of yore, dying to be born a fighter, but to die again once more.'

According to the theory of reincarnation, we may also join up with old friends, or even old foes to continue our outstanding karma.

Have you, for instance, ever taken an instant like, or dislike to someone you have just met for no apparent reason? Have you ever experienced love at first sight, or met someone and felt like you'd known them forever? In some cases, you may be meeting people you have known in other lives. Alternatively, have you perhaps, for no apparent reason, felt really magnetised to a particular place, or felt an intimate recognition, even though you've never been there before, or perhaps a particular time in history that you are particularly fascinated with or drawn to? Do you have any particular birthmarks, likes or dislikes, fears or phobias you cannot explain? These too, may be unconscious memories of past life experiences.

How is it possible to reincarnate? Surely we only have one life?

The body you are in now, yes, it only has one life. We know at physical death that this body will turn to dust. Of course, the body cannot reincarnate because the body is not eternal. It is temporary. It is physical and mortal. The soul within your body however, is eternal and indestructible, and it will leave your body at death.

As we have already seen, a ball of light exists within each one of us that we shall call the soul. At death, this ball of light, the soul, the totality of our consciousness, memory and experience, leaves the physical body and returns to its original source as we saw in the near-death-experience.

But we know that energy does not die! It cannot stagnate! The energy ball, or spirit of the deceased will, like the universe itself, die and be reborn many times. The cycle of creation, the dynamic relationship between energy and matter will continue, so far as we know forever. Energy does not materialise into form and then disintegrate or become disinterested. Energy materialises as form, the form ceases to be, and the energy leaves the form. But energy does not simply retire after creating one form. It continues making and recreating form forever.

It is the same with the energy, the spirit, which has created our forms. When the energy finally leaves the physical body, it will return to the energy source and at some point create another form. The individual soul, or spirit, will then select a new life for itself and choose its future parents ready for incarnation at the relevant moment in time, for the relevant karmic reasons.

Of course, the idea of reincarnation is absolutely absurd if you think that you are your physical body. We know that the physical body will die, and disintegrate.

We cannot reincarnate into the same body again. But the 'soul' which is pure light, consciousness, or energy, can and will recreate itself many times over and produce as many different physical bodies as necessary for its continuing soul evolution here on Earth.

However, as scientists are now discovering, this is not the only planet in the universe. British scientists have now confirmed the existence of other planetary systems. It may be that once you have finished your schooling here on Earth, that you might be promoted, or demoted, to a different planetary system with entirely different cultures and life forms to our own.

According to the Theory of Reincarnation, however, we must incarnate time and time again on the Earth plane until we have perfected and purified the soul, and re-entered the 'Kingdom of Heaven'. Until such time, we must reincarnate back on Earth time after time, life after life, to fulfil our karmic destinies and debts - whatever they may be.

The belief in reincarnation is inherent in many of the world's primal religions, as well as the Eastern philosophies of Hinduism and Buddhism. If it exists in the Western Monotheistic religions, then it is often hidden, deleted or ignored.

'Reincarnation' itself is actually a latin word meaning 'entering the flesh again' and as far back as the 1st century BC, Alexander Cornelius Polyhistor wrote that 'the Gauls teach that the souls of men are immortal, and that after a fixed number of years they will enter a new body'.

Julius Caesar wrote of the Celts in his De Bello Gallico that 'the principal point of their doctrine is that the soul does not die and that after death it passes from one body into another, a firm belief in the indestructibility of the human soul which merely passes at death from one body to another, for by such doctrine alone, they say, which robs death of all it terrors can the highest form of human courage be developed.'

In Africa, the Yoruba believe that reincarnation takes place solely within the family. If they recognise a transcendent soul, or 'Ori', it would be given the special name Babatunde, meaning that 'father has returned', and for girls, Lyatunde meaning 'mother has returned.'

The concept of reincarnating back into the immediate family is not unknown in the theory of reincarnation, but individuals may also choose to reincarnate in soul groups, as well as continuing intense personal relationships with old 'soul mates'. These continuing relationships may be for a specific purpose, either individual or collective, and are a direct result of individual and group karma.

The theory of Reincarnation can be seen to be present in the majority of the world's religions since the dawn of civilization and is so intrinsic in human culture that it predates recorded history. The concepts of 'judgement', 'soul' and 'afterlife' are also recurring themes, and the many modern day testaments of millions of ordinary people from all over the world who have had past life memory or experiences, must surely add to a body of evidence in favour of Reincarnation, far too great to ignore.

We may have been rich in a past life, and now come back to understand the vagaries of poverty. We may have been paupers in one life, now returned to riches and luxury. We may have lived a life of monk-like deprivation, only to return a glutton. We may have been a courtesan and now reincarnated as a nun. We may have been King or Queen, only to return as a slave.

But is there any valid experimental data that could help prove this theory? It would certainly appear that certain factors in the stories of reincarnation and past life memory are without logical or scientific explanation, and incredibly historically accurate in the detail of the memory itself.

In two cases of childhood past life memory which occurred in India, one of which was witnessed and filmed by a BBC TV crew, children who had remembered 'other lives' at an early age, and who had never left their own towns, had accurately described and identified their old homes and families, hundreds of miles away! They were also accurately able to describe their own deaths in the past life, and the bodily part that was injured. In both cases the facts were proved 100% accurate, and both children had birthmarks on the sight of the previous life death injury.

At the University of Virginia in Charlottesville, USA, Professor Ian Stevenson has been investigating over thirty years of research on people who claim to remember past lives. Early in his investigations Professor Stevenson became aware that some who remember past lives had birthmarks or birth defects that corresponded to wounds, usually fatal, on the person whose life was remembered.

His work suggested surprising answers to such questions as 'Why does someone born with a birth defect have that, and not another defect? 'Why do some children in early infancy show phobias when they have had no relevant traumatic experience or model for the phobia in their family'? Indeed, why do we inherit a particular set of circumstances and not another? Why is one soul born into poverty, and another is born with a silver spoon in their mouth?

Okay! I can hear all the scientists and 'nature' Genetic theorists yelling 'It's all in the genes!' Yes, we know that, but do you remember that celestial phenomena like Eclipses, and planetary sound can have a direct effect on genetic material? Yes, and I can hear all the 'nurture' theorists yelling 'It's all in the nurture'! Yes, we know that, but do you remember that celestial and planetary influences can directly affect our psychology and our biology and even the evolution of DNA?

The Birth Chart is a map of the heavenly influences on the genetic material and make-up and can be seen as a cosmic blueprint of life pattern for the incoming soul. The incarnating soul has now chosen its new life in accordance with its karmic destiny. Of course, we generally have little, if any conscious memory of our previous lives and this may be because at the moment of physical birth, literally tens of billions of brain cells actually die. Perhaps these are the cells that carry the memories of past lives. In which case you might argue, is it wrong to know anything more than what we do about our past lives? Of course not!

If a past life has come up in your conscious mind, it is because it is trying to help you to unblock something that has been bothering you, unconsciously for a long time. Maybe in a past life you suffered some trauma that is still echoing in your present life, and indeed in some cases, making your present life intolerable or unhappy? These are called 'past-life blockages' when a traumatic past life experience is disturbing your conscious mind, or interfering with your relationships, or reducing your confidence, killing your ambition, or whatever.

Do you have some form of relationship or life problem you cannot solve? Are you stuck in a 'past life rut'? Perhaps you were left in dire circumstances by a past life-mother, or husband, and you now have an 'irrational' fear of abandonment. Perhaps you drowned and you now have a fear of water. Perhaps you were killed by a snake and now have an instinctive fear of snakes. Perhaps you died in a fire and are now absolutely terrified of it. Perhaps you died in childbirth and are now too frightened to have children. The scenarios are endless.

Whilst children's past life memories seem to occur quite spontaneously between the ages of 2 - 7, for adults it usually requires some form of regression, either through hypnosis or other spiritual practices to access past life memories.

In some cases, adult past life recall or regression, and the subsequent release of a hidden traumatic memory can have some remarkable overall healing effects and allows the release of unconscious fears and phobias.

One case history, that of a woman who had terrifying nightmares of being chased and eaten by a lion night after night for years on end, saw them end instantly after just one session with a hypnotherapist who simply suggested this might have been a past life memory and a past life experience. From that day on, the woman never had another nightmare. The conscious mind had recognised an unconscious past life memory, and the 'irrational' fear and the nightmares were resolved once and for all.

Past Life Regression

'All the world's a stage, and all the men and women merely players.'

William Shakespeare. 'As you Like it'.

For Liz Howard, a housewife and mother working as a Scientist at ICI in Cheshire England, a vivid and haunting dream about an old Tudor manor house was to be the start an unexpected series of events which eventually took her back, under hypnosis, to a former life as Elizabeth Fitton. Born at Gawsworth Hall in 1503, and later a handmaiden to Anne Boleyn, wife of King Henry VIII, Liz was able to describe in detail original features of both the local parish church, and of Gawsworth Hall (known only to the current owners).

The dream Liz had had, had been so intense that 'I could remember it in technicolour detail - so much so, for days and weeks after, this dream kept coming back - it seemed to literally haunt me.'

Six months later and on the way to stay with friends for the weekend, Liz and her family suddenly decided to break their journey at Littlecote, and, by strange coincidence there was a full dress Elizabethan pageant being performed at Littlecote manor that very day.

'Brian - stop the car. That's it! That's the house in my dream!' Both amazed and shocked, Liz and her family got out of the car and walked towards the Manor House. 'I've got this really weird feeling' Liz had told her husband. 'It was up there, back of the house. That's where it happened.' 'How do you know, have you been here before Liz?' asked Brian. 'Not that I know, not that I can think of. I don't know why I know so much about this house.'

Taking the guided tour around the 15th Century Littlecote Manor House, Liz felt nothing more until they got upstairs and went into a small room, off the top landing. Suddenly although it was sunny outside, Liz had the 'impression' that it was dark, windy and rainy. Liz Howard suddenly began to feel visibly upset, when suddenly the tour guide informed them 'This landing is supposed to be haunted - because hundreds of years ago a father killed his own newborn child in this very room.'

Liz and her husband were absolutely astounded by what was happening. Both were extremely sceptical and neither of them knew what to make of it, and yet they were both too intrigued to ignore it. They eventually decided to follow up the case, firstly researching at the local library and finally writing to Joe Keeton, a well known and respected hypnotist with experience in past life regression, who agreed to regress Liz in the hope of finding some answers to something that was disturbing her so greatly and so obviously. The following is an extract from the taped sessions.

Joe: 'Where are you Liz?'
Liz: 'It's cold'.
Joe: 'Where are you?'
Liz: 'In the Tower.'
Joe: 'What tower?'

Liz: 'The Tower of London.'
Joe: 'What year is it?'
Liz: '1536'
Joe: 'What are you doing?'
Liz: (voice becomes distressed) 'I be wi' Lady Anne - she's waiting for the executioner.'

(Liz remembered being one of the four ladies-in-waiting to Anne Boleyn. However, only three names are known in historical accounts and hers is not amongst them. Mysteriously however, the fourth one is actually missing from the records.)

In the next session:

Joe: 'What's your name?'
Liz: 'Tis Elizabeth, I be named after the old Queen.'
Joe: 'What year were you born?'
Liz: '1503'
(NB: Henry VII was King at the time and his wife's name was Queen Elizabeth.)
Joe: 'What year is it now?'
Liz: '1514'
Joe: 'Where do you live?'
Liz: 'Gasworth Hall.'

Liz then proceeded, under hypnosis, to describe in detail Gasworth Hall itself, as well as the rest of her family who had lived there in her past life. In the meantime, the current owners of Gasworth Hall, Tim and Liz Richards, were called in to test Liz on her knowledge of the old buildings.

Joe: 'Where are you Liz?'
Liz: 'I'm in the kitchen.'
Joe: 'What's on the floor Liz?'
Liz: 'Carriage stone - 'tis carriage stone.'

Tim and Liz Richards confirmed that before the floor was replaced with wood in Victorian times, it had indeed been made of carriage stone. This information was only known to the current owners and could not have been in any other way available. The session continued:

Joe: 'What colour is Gasworth Hall Liz?'
Liz: 'Yellow and Black'.

In Tudor times the buildings were painted yellow and black, not white and black as we might assume and Liz and Tim Richards confirmed that no one would have had access to this information at that time.

Tim and Liz Richards were so impressed with Liz's story that they later invited her to Gasworth Hall where she made a visit to the local Church, which she had also described under hypnosis. 'That's odd, it feels different - it feels like it's back to front! The door should be over there!' Liz commented as she and the Richards entered the old Church. Had Liz Howard gotten one fact wrong?

'Of course, in your day Liz', said the vicar as he entered shortly after, 'the Church would have been the other way round. The original entrance was bricked over a hundred years ago.' Tim and Liz Richards were extremely impressed with Liz's story, because no one other than themselves had access to the detailed information Liz Howard seemed to be so familiar with.

Liz and her husband then decided to check the records at Gasworth Hall, and although Liz Howard was keen to find her story was not going to be confirmed by records - still wanting it all to go away and be an elaborate fantasy - they then found all the details of her family and herself, and her life as Elizabeth Fitton at Gasworth. All the details were exactly as Liz Howard revealed under hypnosis.

In her final regression session, Liz found herself at the scene of the baby's murder, in the room at Littlecote.

Joe: 'How old are you now Liz?'
Liz: 'S - Seventy.'
Joe: 'Where are you?'
Liz: 'Littlecote .'
Joe: 'What do you do?'
Liz: 'I work in the kitchens.'
Joe: 'Are you happy?'
Liz: 'Old - pains.'
Joe: 'Come forward 2 years.'
Liz : 'Cold - raining - the baby's come - the midwife has delivered mistress of a boy - (voice gets extremely distressed) he's got no clothes - no- no - no he's going to kill the baby - no I beg you master!'

Liz became so distressed at this point, that Joe Keeton pulled her out of her regression.

'Now I understand why that dream haunted me' Liz later said and perhaps, indeed, it had haunted her for many hundreds of years and many lives until she released that painful and horrific memory. Although Liz Howard later became increasingly sceptical of her own experiences, there is no rational explanation for how she knew about these details - except if she had lived this life hundreds of years in the past.

Although Liz Howard still remains sceptical about her own experiences, one thing still intrigues her - Fitton is also Liz Howard's maiden name in this life.

Commenting on her experiences with hypnosis, however, Liz said that although she never lost sight of who and where she was when she was actually under hypnosis, she seemed to be able to view, with detachment, her emotional past life experiences, as though it was happening in another part of her brain.

Liz Howard's story is by no means unique, and the number of people who are having verifiable past life experience is growing daily.

In her book 'Yesterday's Children', Jenny Cockell describes vivid childhood memories of a past life, in which she was 'Mary', a poor Irishwoman who died at the age of 33, leaving six young children behind her. Her past life death trauma of leaving them was so great that she vowed to return, and was incarnated as Jenny Cockell ten years later.

In her childhood she was obsessed with these memories and drew detailed maps and diagrams of where she used to live. She also gave the name of her then husband and the story of his life in the army. By her teens, Jenny had found records of her life as 'Mary', the town in which she lived - all of which was identical to her childhood memories - and was finally reunited with all her six children.

This kind of vital evidence in favour of reincarnation is by no means unique, and there are many accounts of past life memory which have been reported and investigated which appear to offer valid evidence and information about reincarnation by giving detailed accounts of events and other historical data that could not have been obtained under any other circumstances.

Many children seem to have past life memories quite spontaneously in normal waking consciousness, whereas many adult cases of past life regression usually occur under hypnosis.

One of the first investigators into past life regression was Arthur Bloxham, a hypnotherapist with a keen interest in past life research. In all, he conducted over 400 taped past life regressions. They have become known as the Bloxham tapes.

One of the most intriguing of these, however, occurred to a woman called Jane Evans. Under hypnosis Jane, who had originally consulted Arthur for medical reasons, recalled at least seven of her past lives. The most interesting of these was a past life Jane had remembered as Rebecca, a young Jewish mother in 12th Century York.

Under hypnosis Jane remembered being the wife of a

Jewish moneylender called Joseph. They had two children. She described her life in detail. She was living in the year 1189 just before the Third Crusade when Christian religious fervour was at its worst, and anti-Semitism was widespread.

Reluctant to leave their home, they were eventually forced to flee when an angry mob started to attack the Jewish quarter.

The family fled, keeping the mob at bay by throwing gold coins at them, and eventually took refuge in a Church. They tied up the priest and then hid in the crypt beneath the floor. Eventually, things seemed to quieten down, and Joseph, along with his son, left the church to find food leaving Rebecca and her daughter alone. Their fates were sealed. Shortly after, Rebecca and her daughter Rachel died in the crypt at the hands of an angry mob.

To try and substantiate the mass of historical detail, this case was investigated by Professor Barrie Dobson, an authority on Jewish history at York University, to see if he could substantiate Jane's past life memory. Although he found her story accurate in many details, one thing could not be explained. None of York's mediaeval churches, except for York Minster, had crypts.

However, several months later, in the Spring of 1975, workmen renovating the church of St Mary's Castlegate - which is the church Jane had described under hypnosis - found a secret entrance into the crypt - exactly where she had described it. According to architectural evidence, the crypt appeared to be Norman, or even Roman, in which case, it certainly pre-dated the massacre of 1189 and would have been there at the time, exactly as Jane Evans had described it. No living person could have known about the crypt, other than the woman who had once hidden there in fear of her life.

These accurate historical details defy conventional explanations, such as 'she must have read about it and forgotten it' or 'she must have seen it on the TV and forgotten about it' and so on.

According to psychologist Susan Blackmore, who is obviously as sceptical about reincarnation as she is about everything else, says, and I quote, 'perhaps it's not a reincarnation phenomena at all, perhaps we just underestimate how clever the mind is'. This, I believe, is her way of saying 'she must have read about it and forgotten it' or 'she must have seen it on TV and forgotten about it'.

However, this argument just does not hold water! For instance, how could Liz Howard have possibly 'remembered' the floor was carriage stone. No one ever knew except two people she had never met before. It would have been absolutely impossible.

And how did Jane know about the secret door into the crypt when it had been sealed up for hundreds of years? There are just too many cases, too many accurate historical facts that have been revealed under hypnosis that would be impossible for any living person to know about.

As to the genetic memory theory, I certainly do not give much credence to that and there is absolutely no evidence in its favour. No doubt we have a universal DNA memory, which stores world history from the year dot, in every cell in our body, but I do not believe these past life regressions are dips into a collective memory bank - they are dips into a private memory bank - and so far as I am concerned, dipping into the collective memory would suggest an even stranger phenomena, that of being able to consciously tap directly into DNA collective memory codes.

If past life regression was random, furthermore, you would not expect to have the same past life memory or scenario repeated each time you dipped into the DNA collective memory banks - they would be entirely random, and have little marked effect on your emotional frame of reference.

In many cases of past life regression, the facts are constant. The reactions are constant, as in the case of Liz Howard who had over half a dozen hypnotic regressions, each time going back again and again to exactly the same life scenario. Liz Howard did not randomly access other personalities or DNA collective memories. Her memories were entirely unique and focused. If they were not, then there must be someone else out there who is equally capable of tapping into Elizabeth Fitton's life in Tudor England. Any volunteers?

Furthermore, if we were able to dip into the DNA memory of, say, Aunty Flossie who lived in a cave 500,000 years ago, it's not necessarily going to have the same emotional impact and cause such intense emotional pain as the past life memory seems to evoke - because a random DNA memory it is not a personal experience. The past life experience is so often full of obvious emotional reactions for the subject, as they re-live painful episodes from their past.

The late pioneering regression hypnotherapist Joe Keeton, met journalist Ray Bryant when Bryant was researching a story on past life regression and reincarnation for the Reading Evening Post. Bryant contacted Joe to try regression hypnosis for himself as part of his research.

When Ray was first hypnotised, he immediately regressed to a former life as a soldier serving in the Crimea. He was able to give his first name, Reuben, but could only remember the first three letters of his second name. However, he described his former life in vivid detail.

This was the first glimpse of a former life Ray had apparently had as Rueben Stafford, who was born in Brighton in 1820's and who had moved to Ormskirk Lancashire when he was very young. He became a soldier in the 47th Regiment of Foot. He became a sergeant and saw service in the Crimea, and was married to a woman called Mary.

Andrew Selby, a researcher for Joe Keeton, was already researching at the Records Office at Kew, when he accidentally found a book listing all the casualties in the Crimean War.

Amazingly, he found the records for Sergeant Rueben Stafford, and tested Ray under hypnosis with historical data he had obtained in his research. Ray was able to give every correct detail of every event that was available from Andrew Selby's research - including his own eventual death by drowning. (He committed suicide, old, alone, his wife dead, and in despair he walked into the Thames.)

One interesting fact was that Ray, as Rueben Stafford had suffered a particular wound on a particular day during his time in the Army. The records clearly indicated it. The records read 'slight wound of left hand. Battle of the Quarries 7th June 1855'. They tested Ray: Joe: 'Go back to the time you were in the quarries.' Ray suddenly screamed and clutched his left hand and was so distressed he was immediately taken to another point in time.

As in the case of Liz Howard, Joe Keeton checks his cases with historical records, and finds them all to be accurate to an incredible degree. Ray Bryant went on to experience two further lives under hypnosis.

Were they cheating, had they already seen these historical records?

This certainly does not account for how these people knew about things not contained in local records, like the carriage stone at Gasworth Hall, and the secret chamber in a Yorkshire church, or other precise historical data which none of them would have had access to at any time in their lives, either consciously, or unconsciously.

As with the case of journalist Ray Bryant, many people report having multiple past life memories, which would confirm that we do not just have this life, or even one previous life, but many. The time between one death and a new life is relative to each individual.

Often there are clues at the beginning of this life you are having now, as to your own past lives, and often there will be distinct echoes and reflection of the previous life, immediately prior to the current one. For instance, you might have been a sailor in another life, and find in your youth that you live near a port, or harbour. You might have been a musician in an earlier life, and take up music at an early age with ready-made talent.

Like many other celebrities, former Genesis frontman Phil Collins is a modern day exponent of reincarnation, and firmly believes he was incarnated at The Alamo as Texan courier John W Smith, a man who was known as El Colorado - the redhead because of his hair. It all started when he was a young boy. Collins says: 'I just had a total and very early fascination with the Wild West. It just got me, and never left me.' He would ask his parents to buy him raccoon-skin hats and toy rifles as birthday presents. His fascination with that period in history continues to this day.

Do faint echoes of our former lives pervade the current life in some form or another? Mozart was a genius at 4. Did he have past lives where he had already learned the rudiments of music, and did General Patton, who was fascinated by Napoleon, and who believed passionately about reincarnation, live once as the great Bonaparte himself? Does this account for his exceptional skill as a warrior?

The soul has many layers, like an onion, each life is part of our totality, and we may carry our gifts and talents, latent within us, as we do our hopes and fears. From lifetime to life time we traverse time and space reaping the rewards of our own proportional existence. We are born, and we die, to be reborn again until we reach perfection.

In Hinduism it is believed that God, who is One and the source of all life, creates the material universe from itself and yet is simultaneously the universe itself. This journey, from spirit to matter, and back again is also described in detail in the Kabbalistic Tree of Life. God, Light, Consciousness, call it what you will, creates the material universe, and is omnipresent within it simultaneously.

As with all cycles, the cycle must be complete and the

energy must return to its source. From life to life we go through all the tests and lessons that we have set ourselves until we understand the nature of illusion, the illusion of physicality, and return to our spiritual source.

The Theory of Reincarnation states that all our previous actions will result in reactions, which we ourselves have already set in motion in previous lives and until we reach perfection we must repeatedly reincarnate on Earth to learn our karmic lessons.

Our Karma, that which we set in motion in previous lives and previous times, is still unfolding and will continue its cause and effect scenario until balance is restored and the lessons learned.

Until then the original action still exists in time and space as do all the reactions to it.

Return to the Source.

'And whoever, at the time of death, quits his body remembering Me alone, at once attains My nature. Of this there is no doubt. He who meditates on the Supreme Personality of Godhead, his mind constantly engaged in remembering Me, undeviated from the path, he oh Partha (Arjuna) is sure to reach Me.'

Baghavad Gita 8.5 - 8.8

And so the drop, having been evaporated from the ocean, finally returns to its source, its home. The eternal interchange of energy has taken us on our journey from energy to matter, and now we must return from whence we originally came, matter must return to energy.

Everything is Energy. Everything is Energy made manifest. Nothing exists without energy, not you, not me, not the cosmos, not anything.

It is within and without simultaneously.

This Energy furthermore, capable of creating infinite worlds, is the beginning, the middle and the end. The Alpha and Omega. Nothing comes into being without this energy. No physical form can exist without it, it is the **G**enerator, **O**perator and **D**estroyer of worlds. It precedes material creation. It is the indisputable source of all life. The physical form does not exist without it. The form - whether it be large or small, animal, vegetable or mineral - is a direct manifestation of this energy.

What is this energy that has the power to create infinite worlds?

Is this Energy in fact, God?

Have we anthropomorphised G.O.D to such a degree that we may not have considered that G.O.D *is* this energy and that if this energy *is* God, then omnipresence is a quantum as well as spiritual reality?

Or, are we, perhaps, expecting to see a God who resembles us in some or any way? Do we create a concept of God based on our own limited imaginations and fragile emotions?

What if God *is* this endless, eternal, infinite Energy? If so, then this Energy now becomes divine and we come to understand everything is One. It is the One made visible. Everything is connected.

Every living being, be they human, animal or otherwise, is this divine energy made flesh and contains this divine energy within itself - this energy is conscious, is consciousness itself - pure mind - and never dies, merely changing forms in an endless cycle of birth, death and rebirth.

The individual soul is the portion of energy that pre-exists the physical self, and is the causative factor behind its creation. The soul energy of the individual was present before material creation.

The individual soul is a drop and came from the ocean of God and to God it will return.

The Soul is the part of us which is eternal, indestructible and infinite.

It will never die, not even when the body dies.

To try to understand and incorporate the reality of the interconnectedness, of the Oneness of all things, will surely help us evolve as beings worthy of the stewardship of this planet and ensure that we understand our place and purpose in the cosmos and to treat this divine creation - in all its myriad manifestations - with the utmost respect. For whatever we do, will come back to us. That is the Law. That is the law of Karma. The Law of Cause and Effect.

We are on the cusp of great changes in both civilisation and culture, and it is only, I believe, by recognising and respecting the divine energy within ourselves, and all of creation that we can move forward with renewed understanding and respect for each other, the Earth, and all the other amazing, irreplaceable creatures that we share this truly unique and amazing planet with.

There is no escape, there is no Plan B, there is no Planet B. If we do not amend our ways, each and everyone of us, we will continue on our path to the next mass extinction. It is only by recognising our interconnectedness, and understanding that everything and everyone contains this incredible, unfathomable, holy and divine energy, that we can progress in the right direction and save ourselves from self destruction as a species and move forward to a shared and better future.

We are all floating together on this little planet called Earth, each one of us together, like petals on a flower, mankind on the Earth, a globe, relative in size, vast compared to the spinning atom, minute compared to the spinning Universe.

Truly, we are star children one and all - and we are all One.

Moreover, one last question must remain. When we return to the source of all things, do we stay there forever, or are we ejected out again in some unfathomable time in the future, to commence the cycle all over again, once again? Universe after universe, life after life, aeon after aeon?

Ah, sweet mystery!

BIBLIOGRAPHY.

Hitlers Pope, The Secret History of Pius XII
John Cornwell. Viking. London 1999.

Gone with the Wind in the Vatican
1 Millenaria. Kaos Edizioni. Italy 1999.

The Hermetica - The Lost Wisdom of the Pharaohs
Timothy Freke and Peter Gandy. Piatkus Books. London. 1997

Hyperspace
Michio Kaku. Oxford University Press. UK 1994.

Hall of the Gods
Nigel Appleby. Heinemann. UK 1998

The Tenth Dimension
Michio Kaku. (out of print)

Superforce
Paul Davis. Penguin. London 1984

The Physics of Immortality
Frank Tipler. Pan Books. London. 1994.

The Music of Life
Hazrat Inayat Kahn. Omega Publications. NY. 1998.

The Elegant Universe.
Brian Greene. Jonathan Cape. London. 1999.

Christ and The Cosmos
Professor E.H. Andrews. Evangelical Press. UK. 1986.

The Da Vinci Code
Dan Brown. Doubleday Books. 2003

The Sun of God
Gregory Sams. Weiser Books. 2009

The Timaeus
Plato. Hackett Publishing Co. USA. 1999

The Round Art
A.T. Mann. Paper Tiger UK 1979.

The Book of the Eclipse
David Ovason. Arrow. London. 1999.

The Secret Messages in Water
Dr Masaru Emoto. Beyond Words Publishing. 2004

Cosmic Influences on Human Behaviour
Michel Gauquilan. Garnstone Press. London 1973.

The Secret Life of Plants
Peter Tompkins and Christopher Bird. Allen Lane. London. 1974

Hands of Light
Barbara Ann Brennan. Bantam Books. NY. 1988.

The Dancing Wu Li Masters
Gary Zukav. Quill. NY. 1979.

The One: How an Ancient Idea Holds the Future of Physics.
Heinrich Pas. Audio book. 2023

The Hypnotic World of Paul McKenna
Paul McKenna. Faber and Faber. London. l993.

The Doors of Perception
Aldous Huxley. Flamingo Books. (Harper Collins) London l994.

Food of the Gods
Terence McKenna. Rider. London. l992.

The Teachings of Don Juan
Carlos Castaneda. Penguin. l970.

Autobiography of a Yogi
Paramahansa Yogananda. Self Realisation Fellowship. LA. l975.

Creative Visualisation
Shakti Gawain. Bantam. N.Y. l982.

The God Experiment
Russell Stannard. Faber and Faber. London. l999

Mind Sculpture: Your Brain's Untapped Potential.
Ian Robertson. Bantam. London. l999.

The Tao of Physics
Fritjof Kapra. Flamingo. London. l983.

Dogs That Know When Their Owners are Coming Home and Other Unexplained Powers of animals.
Rupert Sheldrake. Hutchinson. London. l999

Prophecy
R.J. Stewart. Element Books. Dorset. l990.

The Bible Code
Michael Drosnin. Orion Paperbacks. UK 1997

The Essential Jung
Anthony Storr. Fontana Press. London. 1998

An Experiment with Time.
J.W. Dunne. Faber and Faber. London 1958.

Peter Pan and Wendy
J.M. Barrie. Pavilion Books. London. 1997.

Reflections on Life after Life
Raymond A. Moody. Bantam. NY. 1978.

Yesterday's Children: The Search for My Children From The Past.
Jenny Cockell. Piatkus Books. UK. 1993.

The Aquarian Gospel of Jesus the Christ.
Eliphas Levi. Fowler. London. 1964.

The Ghost of Flight 401
Fuller. Souvenir Press. UK. 1986

Talking to Heaven.
James Van Praagh. Piatkus. London 1998.

Divine Intervention.
Hazel Courteney. C.I.M.A. Books. London. 1999.

Index

A

Abbess of Lacock 167
Aborigines 108, 142
Acupuncture 66, 72, 77
Africa 51, 70, 77, 92, 212
Akashic Records 174
Alchemy 45
Alker, Bill 100
Amazon 92
Andrews, Professor EH 38, 230
An Experiment with Time 145, 232
Angels 14, 188, 189, 192
Archbishop of Canterbury 15
are Coming Home and Other 126, 231
Aromatherapy 77
Asaf Yefet 99
Asia 92
Astrology 4, 11, 15, 40, 41, 42, 43, 44, 45, 46, 47, 48, 49, 50, 52, 54, 55, 60, 138, 139, 140, 202
Astronomy 42, 45, 58
Aum (Om) 34
Australia 77, 108, 142
Autobiography of a Yogi 111, 231
Azande Tribe 70

B

Bacairis 70
Bardo Thodal, The 172
Barrie, JM 147, 221, 232

Battle of the Quarries 224
Beck, Dr. Robert 73
Becker, Dr. Robert 67
Bell, JS 130
Bennett 81, 82
Benson, Dr Herbert 118
Bergson 113
Bhagavad Gita 148, 183, 206
Bible Code, The 138, 139, 232
Bible, The 22, 26, 36, 49, 50, 61, 114, 138, 139, 140, 164, 196, 198, 232
Bird, Christopher 64, 230
Bird, R.C. 118
Birth chart 49, 54, 55, 214
Blackmore, Dr. Susan 46, 47, 158, 222
Blake, William 50, 90, 96
Bloxham, Arthur 220
Boleyn, Anne 10, 215, 217
Book of the eclipse, The 56, 230
Bragg, Melvyn 151
Brahma 33, 39
Brahmins 50, 136
Britain 11, 12, 44, 47, 48, 49, 50, 101, 139, 176, 186
Britannic, The 118
British School of Hypnosis 88
Broad, Dr CD 113
Bryant, Ray 223, 224
Buddha 13, 70, 110
Buddhism 13, 23, 83, 107, 168, 196, 211

C

Caesar, Julius 211
Cainer, Jonathan 49
Cannabis 98, 99, 100, 101, 102, 153
Capra, Fritzjof 124

Cardinal Hume 166
Carey, Dr George 15
Castaneda, Carlos 96, 231
Cayce, Edgar 174
Celts 115, 211
Chakras 27, 73, 75, 76
Chalmers, David 208
Chanting 104, 105, 107, 108
Ch'i 72
China 77, 127
Christ and the Cosmos 38, 230
Christianity 9, 12, 13, 23, 26, 50, 83, 104, 110, 197
Church of England 10, 11, 15
CIA (Central Intelligence Agency) 98, 134
Clark, David 94
Clarke, Alan 155
Cleopatra 123
Coady, Bernadine 88
Cockell, Jenny 220, 232
Confucianism 23
Cornwell, John 10, 11, 229
Council of Nicea 14, 26, 197
Courteney, Hazel 191, 232
Cowan, Clyde 30
Crazy Horse 142
Crimean War 223
Culpepper, Sir Thomas 100

D

Dalai Lama 168, 196
Dancing Wu Li Masters, The 78, 230
Davey, Dianne 188
Davidson, Verena 73
Davis, Professor Paul 32, 131, 229
Dee, Dr John 47

Diana, Princess of Wales 118, 191, 192
Divine Intervention 191, 232
DNA 56, 103, 214, 222, 223
Dobson, Professor Barrie 221
Dogon Tribe 51
Dogs That Know When Their Owners 126, 231
Don Juan 96, 97, 231
Doors of Perception, The 94, 96, 112, 231
Dowding, Sir Hugh 186, 187, 209
Dreams 4, 32, 92, 108, 132, 141, 142, 143, 144, 145, 146, 147, 165, 176, 190, 199
Drosnin, Michael 138, 232
Druids 44, 50, 64
Dunne, WG 145, 232

E

Eastern Airlines 181, 182
Eckhart, Meister 129
Eclipse 56, 230
Ecstasy 95, 102, 103, 106, 170
Egypt 70, 77, 123, 167
Eicke, Paul 163
Einstein, Albert 25, 38, 122, 130, 131, 156
Elegant Universe, The 36, 229
Emerald Table, The 45
Emperor Constantine 26
England 10, 11, 15, 43, 46, 47, 50, 88, 104, 118, 167, 215, 223
Ephemeris 55
ESP 4, 9, 15, 129, 133, 134, 137
Essential Jung, The 141, 232
Europe 10, 48, 97, 101, 209
Evans, Jane 220, 221

F

Fenwick, Dr. Peter 158
Fitton, Elizabeth 215, 218, 219, 223
Food of the Gods 95, 231
Ford, Paul 151
Fort Mead 134
Franco 11
Frank Tipler 32, 229
Freilich, Jenny, 182
Furnham, Professor Adrian 45

G

Gates, Bill 11, 117
Gauquelin, Michael 58, 59
Gawain, Shakti 116, 231
Gawsworth Hall 215
Genesis 22, 49, 61, 149, 225
George and Pilgrim, The 179, 180
Gerard, Robert 74
Ghost 167, 173, 174, 175, 176, 177, 178, 180, 181, 182, 183, 187, 192, 232
Ghost of Flight 401 232
Ghosts 5, 135, 172, 175, 176, 177, 179, 181, 182, 183, 187, 190
Gill, Charles 149
Glastonbury 43, 50, 179, 180
Glastonbury Abbey 179, 180
Glastonbury Tor 180
God Experiment, The 117, 231
Gone with the Wind in the Vatican 11, 229
Greene, Brian 36, 229
Green, Michael 30, 50, 64, 74, 76, 143

H

Hale-Bopp Comet 138
Hall of the Gods 31, 229
Hameroff, Dr Stuart 164, 165, 208, 209
Handel 35
Hans, Dr Jenny 34, 79
Harary, Dr Keith 134
Hashomer Hospital 99, 100
Hasidic Jews 197
Hatch, Donald 35
Hazrat, Inayat Khan 35, 229
Healing 4, 15, 49, 72, 73, 74, 76, 77, 86, 105, 215
Heaven 9, 11, 14, 18, 22, 57, 80, 90, 97, 108, 110, 111, 113, 160, 164, 165, 170, 171, 172, 184, 189, 198, 204, 208, 211, 232
Hell 9, 14, 170, 171
Hemp 100
Herbalism 49
Hermetica, The 17, 59, 61, 229
Heroin 102, 185
Highfield, Dr Roger 45
Himis 50
Hinduism 13, 23, 39, 83, 196, 211, 225
Hitler, Adolf 10, 199
Hoffman, Dr Albert 98
Howard, Liz 215, 216, 218, 219, 222, 223, 224
Huxley, Aldous 94, 103, 231
Hyperspace 30, 229
Hypnosis 4, 84, 85, 86, 88, 89, 120, 175, 207, 214, 215, 217, 218, 219, 220, 221, 222, 223, 224
Hypnotherapy 88
Hypnotic World of Paul McKenna, The 85, 231

I

India 50, 70, 77, 115, 136, 138, 213
In Search of Faith 11
Institute of Psychiatry 82, 159
Inyushin, Dr Victor 68
Islam 11, 13, 83, 104, 197
ITV 154

J

Jenny, Hans 34, 35, 79, 182, 220, 232
Jerusalem 50
Jesus Christ 13, 80
Jesus College, Cambridge 10
John Radcliffe Hospital 159
Jonas, Dr 55
Joseph of Arimathea 50
Judaism 13, 23, 26, 83, 84, 104, 197
Judeo-Christian 9, 32
Jung, Carl 14, 79, 80, 113, 122, 141, 170, 232

K

Kabbalah, The 26, 27, 31, 32, 33
Kabir 61
Kaku, Michio 30, 32, 229
Kalahari 105
Karma 5, 171, 193, 194, 195, 196, 197, 198, 199, 200, 201, 204, 205, 209, 212, 226, 228
Kazakh University 68
Keeton, Joe 216, 219, 223, 224
Kelly, Edward 47
Kepler, Johannes 58
Kingston Hospital 160
Kirlian Photography 65, 76, 77, 138

Kirlian, Simyon 65
Kirsch, Dr Daniel 83
Krippner 147
Kulagina, Nina 191

L

Laing, RD 98
Lambert, Barbara 161
Lanza, Dr Robert 156
Lao Tse 142
Laurentian University 111
Law of Cause and Effect 5, 21, 112, 193, 194, 195, 196, 197, 198, 199, 204, 228
Lawrence, Stephen 200
Leninger, James 206
Leningrad 66, 191
Levi, Eliphas 232
Littlecote Manor 216
Longespee, William 167
Lord of the Dance 117
LSD 94, 97, 98, 153
Lynch, Sonia 189

M

Macrocosm 19, 28, 61
Madeley, Richard 154
Maimonides Medical Centre 143
Mann, AT 53, 230
Marinelli, Monsignor 11
Marley, Bob 193
Mary Queen of Scots 47
Masaru, Dr Emoto 56, 230
Masood, Zafar 174
Maya 20

McKenna, Paul 85, 231
McKenna, Terence 95, 231
Meditation 4, 12, 13, 83, 87, 99, 105, 106, 107, 109, 110, 111, 113, 118, 124, 149, 154, 158
Mescaline 94, 95
Mesmer, Franz 86
Microcosm 19, 22, 28, 59, 61
Microtubules 165, 208, 209
Mind/Body Medical Institute 118
Mind Sculpture, Your Brain's 120, 231
Mohammed 13
Monotheism 109
Moody, Raymond 162, 232
Moore, Patrick 45
Morris, Professor Robert L 133
Mozart, Wolfgang Amadeus 225
MRI (Magnetic Resonance Imaging) 82, 93
Multiple Sclerosis 100
Mundaka Upanishad 71
Music of Life, The 35, 229
Mussolini 11
Mysticism 12, 13, 14, 17, 106

N

Napoleon, Bonaparte 209, 225
NASA 52, 57
National Head Injury Association 100
Near-Death-Experience 9, 158, 159, 162, 165, 167, 170, 171, 210
Nelson, JH 53, 54
New England Deaconess Hospital 118
New Testament 10, 14, 34, 114, 141, 194, 200
Newton, Sir Issac 21, 46, 47, 193
Newton's Third Law of Motion 21
New York University 31
Nigel Appleby 31, 229

Nirvana 18, 110, 204
North American Indians 142
Nostradamus 139
Nottingham University 77

O

Old Testament 27, 49, 61
Olympic, The 118
Out-of-Body Experience 96, 106, 149, 150, 151, 152, 153, 154, 155, 170, 183
Ovason, David 56, 230

P

Palmer, David 153
Panpsychism 23
Paradise 5, 15, 168, 170, 172
Paramahansa Yogananda 111, 231
Parnell, Chris 163
Patton, General George 209, 225
Penrose, Sir Roger 164, 208
Persinger, Dr Michael 111
Peter Pan 147, 232
PET scanner 120
Physics of Immortality, The 32, 229
Plato 40, 104, 230
Pope Pius XII 10
Pope Urban 11
Prediction 4, 44, 137, 139, 141, 144
Premonition 129, 144
Prophecy 4, 44, 137, 139, 141, 231
Psychics 83, 134, 135, 191
Pyatnitsky, Dr Lev 119
Pyramids 47, 51, 115
Pythagoras 28, 57, 195

Q

Quantum Theory 32, 164, 208
Queen Elizabeth I 47
Queen Victoria 99
Quran 18, 196, 198, 203

R

Rabin, Yitzhak 138
Radin, Dean 125
Ravitz, Dr Leonard 67
RCA 53, 54
Read, Harriet 37, 45, 46, 48, 50, 64, 71, 87, 104, 138, 221, 222, 224
Reading Evening Post 223
Records Office 223
Reflections on Life after Life 162, 232
Reincarnation 5, 11, 15, 18, 70, 86, 171, 187, 195, 196, 197, 198, 199, 204, 206, 207, 208, 209, 211, 212, 213, 220, 222, 223, 225, 226
Reins, Frederick 30
Remote Viewing Programme 135
Reppert, Steven 135
Resurrection 5, 15, 165, 166, 167, 170
Richards, Liz 217, 218
Richards, Tim 217, 218
Rips, Dr Eliyahu 138
Riverdance 117
Robertson, Ian 120, 231
Robinson, Chris 144, 145, 190
Roman Catholic Church 10
Round Art, The 53, 230
Royal College of psychiatrists 158
Russian Academy of Sciences 119

S

Sat Chit Ananda 94
Schufler, Diane 153
Schwarz, John 30
Secret Life of Plants, The 64, 230
Selby, Andrew 223, 224
Sephiroth 27
Sergeyev, Genaday 191
Sharma, Dr Tonmoy 82
Shaywitz 81, 82
Sheldrake, Dr Rupert 68, 126, 127, 128, 152, 231
Shri Yantra 34, 79, 80
Shvetashvatara Upanishad, The 205
Sitting Bull 142
Skull, Derek 160, 161
Socrates 64
Spirit Guides 5, 172, 187, 188, 191
SQUID (Superconducting Quantum Interference Device) 67
Sri Ykteswar 111
Stafford, Reuben 223, 224
Stannard, Russell 118, 231
Steiner, Rudolf 108, 174
Stevenson, Professor Ian 167, 176, 213
Stewart, RJ 137, 231
St. John 33, 77
St. John's Wort 77
St. Luke 168
St Mary's Castlegate 221
St. Michael's Mount 43
Stonehenge 12, 43, 44, 47, 51, 115, 179
St. Paul 149
St. Thomas's Hospital 159
Sufi 35, 61, 108
Superforce 32, 131, 229

Superstring Theory 30, 31, 32, 33, 35, 36, 60, 169
Swann, Inigo 134
Swayze, Patrick 176

T

Tai Chi 39, 81
Talking to Heaven 184, 232
Tampere University Of Technology 115
Taoism 23
Tarot 140
Teachings of Don Juan, The 96, 231
Telepathy 4, 9, 15, 96, 126, 129, 130, 133, 134, 137, 185
Tel Hashomer Hospital 99
Tenth Dimension 32, 229
Thoth 59
Tibet 50
Timaeus, The 40, 230
Titanic, The 118
Tompkins, Peter 64, 230
Tonoscope 34, 58, 79, 80
Transcendental Experience 4, 12, 34, 109, 131
Tree of Life 27, 32, 200, 225
Trismegistus 45, 59
Tzu, Lao 109

U

Unified Field Theory 32
University College London 45
University of Edinburgh 133
University of Massachusetts 135
University of Nevada 125
University of Sheffield 104
University of Southampton 59
University of Tel aviv 99

University of Virginia 167, 176, 208, 213
Untapped Potential 120, 231
Upstate Medical School, New York 67

V

Van Praagh, James 184, 185, 186, 232
Vatican 10, 11, 14, 197, 229
Vedic Philosophy 71
Voyager 56

W

White Star Lines 118
William and Mary University 67
Williams, Jean 159, 160
Witchcraft Act 48
Wordsworth, William 28
World War II 66, 138, 139

Y

Yesterday's Children 220, 232
Yogis 70, 108, 115, 136, 155, 197
York Minster 221
York University 31, 221
Yoruba 212

Z

Zen 135
Zimmerman, Dr John 74
Zodiac 55, 59
Zukav, Gary 78, 230

ABOUT THE AUTHOR.

Crystal Love started her career as a writer in advertising and publishing.

Disillusioned with the superficiality of the commercial world she took a year's sabbatical during which time she had a series of profound mystical and supernatural experiences which were to change her life irrevocably, both personally and professionally.

She subsequently devoted herself to in depth study and research into these supernatural phenomena, determined to find a logical and reasonable explanation. She has previously spoken on TV and Radio discussing these topics and has also had articles in Cosmopolitan Magazine, Sunday Mirror Magazine, Sunday Express and the Daily Express.

She is also a painter and ceramic artist with a keen interest in film and music production.

www.ingramcontent.com/pod-product-compliance
Lightning Source LLC
Chambersburg PA
CBHW071338080526
44587CB00017B/2886